Critical Thinking

for use with

Elementary Statistics
A Step by Step Approach

Fifth Edition

Allan G. Bluman
Community College of Allegheny County

Prepared by
James A. Condor, Ph.D.
Manatee Community College

Boston   Burr Ridge, IL   Dubuque, IA   Madison, WI   New York   San Francisco   St. Louis
Bangkok   Bogotá   Caracas   Kuala Lumpur   Lisbon   London   Madrid   Mexico City
Milan   Montreal   New Delhi   Santiago   Seoul   Singapore   Sydney   Taipei   Toronto

The McGraw·Hill Companies

Critical Thinking Workbook for use with
ELEMENTARY STATISTICS: A STEP BY STEP APPROACH, FIFTH EDITION
ALLAN G. BLUMAN

Published by McGraw-Hill Higher Education, an imprint of The McGraw-Hill Companies, Inc., 1221 Avenue of the Americas, New York, NY 10020. Copyright © The McGraw-Hill Companies, Inc., 2004. All rights reserved.

No part of this publication may be reproduced or distributed in any form or by any means, or stored in a database or retrieval system, without the prior written consent of The McGraw-Hill Companies, Inc., including, but not limited to, network or other electronic storage or transmission, or broadcast for distance learning.

This book is printed on acid-free paper.

1 2 3 4 5 6 7 8 9 0 DCD/DCD 0 9 8 7 6 5 4 3

ISBN 0-07-254913-0

www.mhhe.com

TABLE OF CONTENTS

INTRODUCTION		v
1-1	STOPPING ON A DOLLAR	1
1-2	ATTENDANCE AND GRADES	2
1-3	SAFE TRAVEL	3
1-4	AMERICAN CULTURE AND DRUG ABUSE	4
1-5	JUST A PINCH BETWEEN YOUR CHEEK AND GUM	5
1-8	MORE SMOKE THAN YOU THINK	6
2-1	WHICH GRAPH TO USE	7
2-2	AFTER SCHOOL ACTIVITIES	8
2-3	SELLING REAL ESTATE	9
2-4	LEADING CAUSES OF DEATH	10
2-5	SPORTS DRINKS	11
3-1	COLLEGE TUITION	12
3-2	TEACHER SALARIES	13
3-3	BLOOD PRESSURE	14
3-4	DETERMINING DOSAGES	15
3-5	THE NOISY WORKPLACE	16
3-6	MEN'S FRAGRANCES	17
4-1	PLAYING THE LOTTERY	18
4-2	TOSSING A COIN	19
4-3	WHICH PAIN RELIEVER IS BEST?	20
4-4	GUILTY OR INNOCENT?	21
4-5A	HOW MANY WAYS COULD THERE BE?	22
4-5B	GARAGE DOOR OPENERS	23
4-5C	SEATING ARRANGEMENTS	24
4-6	WHICH METHOD TO USE	25
4-7A	SUBTLE DIFFERENCES	26
4-7B	MISLEADING WORDS	27
5-2A	DRINKING AND DRIVING	28
5-2B	DROPPING COLLEGE COURSES	29
5-3	EXPECTED VALUE	30
5-4	UNSANITARY RESTAURANTS	31
5-5	HOW SAFE ARE YOU?	32
5-6	CHOOSING THE BEST DISTRIBUTION	33
6-1	AVERAGE INCOME	34
6-2	NORMAL HEALTH	35
6-3	THE Z-SCORE	36
6-4	SMART PEOPLE	37
6-5	CENTRAL LIMIT THEOREM	38
6-6	HOW SAFE ARE YOU?	39
6-7	THE PROBLEM WITH NORMAL	40
7-2A	CONTRACTING INFLUENZA	41
7-2B	MAKING DECISIONS WITH CONFIDENCE INTERVALS	42
7-3	SPORT DRINK DECISION	43
7-4	DECEIVING STANDARD ERRORS	44
7-5	FIGHTING DEPRESSION	45

7-6	HAZARDS OF SMOKING	46
8-1	EGGS AND YOUR HEALTH	47
8-2	QUITTING SMOKING	48
8-3	BEE STING THERAPY	49
8-4	MEN'S FRAGRANCES	50
8-5	HAIR CONTROL	51
8-6	TESTING GAS MILEAGE CLAIMS	52
8-7	CONFIDENCE INTERVALS AND HYPOTHESIS TESTING	53
8-8	HOW MUCH NICOTINE IS IN THOSE CIGARETTES?	54
9-1	DOES REPLICATION CONFIRM?	55
9-2	ARTIFICIAL SWEETENERS AND YOUR HEALTH	56
9-3	VARIABILITY AND AUTOMATIC TRANSMISSIONS	57
9-4	TOO LONG ON THE TELEPHONE	58
9-5	MEMORIZING AND FAMILIARITY	59
9-6	JOBS AND PARENTING	60
9-7	REDUCING DRUG ABUSE	61
10-1	SALT AND BLOOD PRESSURE	62
10-2	STOPPING DISTANCES	63
10-3	MATH CONFIRMS IT	64
10-4	STOPPING DISTANCES REVISITED	66
10-5	INTERPRETING SIMPLE LINEAR REGRESSION OUTPUTS	67
10-6	MORE MATH MEANS MORE MONEY	68
10-7	HORSEPOWER AND PRICE	69
11-1	NEW CAR COLORS	73
11-2	NEVER THE SAME AMOUNTS	74
11-3	SATELLITE DISHES IN RESTRICTED AREAS	75
11-4	CATEGORIZING CONTINUOUS DATA	76
12-1	COMPUTER-AIDED INSTRUCTION	77
12-2	ONE BAD APPLE	78
12-3	COLORS THAT MAKE YOU SMARTER	79
12-4	TRADITION OR TECHNOLOGY?	80
12-5	AUTOMOBILE SALES TECHNIQUES	81
13-1	HIGH SCHOOL CRIMES	83
13-2	BETTER SAFE THAN SORRY	84
13-3	NO PAIN NO GAIN?	85
13-4	SIDE EFFECTS FROM MEDICATION	86
13-5	MEMORIZING AND FAMILIARITY	87
13-6	HIGH SCHOOL SUSPENSIONS	88
13-7	DOES MORE MATH MEAN MORE MONEY?	89
13-8	PREDICTING RADIOACTIVE WASTE MOVEMENT	90
14-1	SMOKING BANS AND PROFITS	91
14-2	THE WHITE OR WHEAT BREAD DEBATE	92
14-3	TRADITIONAL FAMILY VALUES	93
14-4	SIMULATIONS	94
14-5	MONTE CARLO METHODS	95
14-6	SURVEYING AMERICANS	96
SOLUTIONS		97

INTRODUCTION

This *Critical Thinking Workbook* is a supplementary activity workbook designed to accompany Allan Bluman's *Elementary Statistics: A Step by Step Approach*, Fifth Edition. There are activities for most sections in the textbook (some sections have two separate activities meant to be assigned together depending on available time). Many activities were pilot-tested in an elementary statistics classroom. Those that were most successful in increasing motivation and understanding make up this workbook. The data and situations in the activities are fictitious, but are based in real-life situations. The activities can be assigned as in-class activities, outside activities, homework, group activities, or used for in-class discussion to supplement lectures.

Many of the activities are marked **(Computer)** which means a computer would greatly facilitate the computations. The computer activities can be used without a computer, however many of the computations can easily be done by hand or with a scientific calculator, but require extended time. No specific statistical computer software is required. The outputs shown in the workbook are generic or they are from MINITAB Release 13 for Windows. Basic statistical calculators or graphing calculators can also be used to accomplish most of the tasks. The focus of the activities is not on developing computational skills. Many exercises in the textbook can be used for that.

The *Critical Thinking Workbook* activities were developed to help students connect concepts that are typically presented in an isolated fashion. They are not meant to be so-called "problem-solving activities." Most recent research, and much frustration, has shown that students cannot make the jump from learning isolated facts to true problem solving. The key breakthrough for elementary-statistics students is to be able to make the connections between isolated facts. Students can then take the next step in the lengthy progression to becoming a true problem solver.

Each of the activities has many questions associated with it. Some of the questions do not have specific answers. They are meant for discussion to generate thought for the students and direction for the instructor. There are also activities for most of the introductory sections to focus thoughts on key concepts in the upcoming sections. The activities for the summary sections tie together the major concepts of the previous sections for each chapter.

James A. Condor, Ph.D.
Department of Mathematics
Manatee Community College

Also available for student purchase to improve your success in this course when using *Elementary Statistics: A Step by Step Approach, 5e* by Allan G. Bluman:

Student Study Guide
By Pat Foard of South Plains College, this study guide will assist students in understanding and reviewing key concepts and preparing for exams. It emphasizes all important concepts contained in each chapter, includes explanations, and provides opportunities for students to test their understanding by completing related exercises and problems.

Student Solutions Manual
By Sally Robinson of South Plains College, this manual contains detailed solutions to all odd-numbered text problems and answers to all quiz questions.

MINITAB Manual
By Gerry Moultine of Northwood University, this manual provides the student with how-to information on data and file management, conducting various statistical analyses, and creating presentation-style graphics while following each text chapter.

TI-83 and TI-83 Plus Graphing Calculator Manual
By Carolyn Meitler of Concordia University Wisconsin, this patient, practical manual teaches students to learn about statistics and solve problems using these calculators while following each text chapter.

Excel Manual for Office 2000
By Renee Goffinet and Virginia Koehler of Spokane Falls Community College, this workbook is specially designed to accompany the textbook and provides additional practice in applying the chapter concepts while using Excel.

Videos
New to this edition are text-specific videos available on VHS and CD-ROM that demonstrate key concepts and worked-out exercises from the text plus tutorials in using the TI-83 Plus Calculator, Excel, and MINITAB, in a dynamic, engaging format.

1-1
STOPPING ON A DOLLAR

Statistics is a numerical communications course. The more you learn about statistics, the better decisions you will make when evaluating communication involving numbers. Evaluate the following results from a study on braking distances of the most popular brands of bicycles and decide which would be the best buy.

Getting More For Your Money

Brand	Price	Wet Braking Distance
Huffy	$110	55 ft
Murray	79	75
Pacific	145	30
Magna	99	62
Euro	175	25

The bikes were tested on smooth, wet pavement. The bikes were new and comparable in type. Stopping distances used for each bike are the best of four stops. The bikes were driven at 15 mph and then the brakes were firmly applied. The bikes were either 24 inch or 26 inch type.

---

1. What are the details of the study?
2. Do you think the results would be different given different conditions?
3. Does the article mislead the consumer in any way?
4. Are the bikes' brakes or the tires being tested?
5. What impression does the article give by listing the prices of the bikes along with the braking distances?

1-2
ATTENDANCE AND GRADES

Read the following article on attendance and grades and answer the questions.

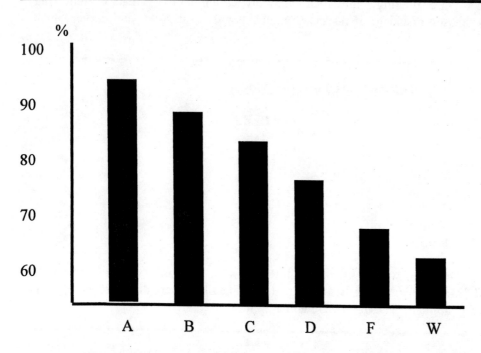

The chart shows the results of a study conducted at Manatee Community College. The data revealed that students who attended class 95 - 100% of the time usually received an A or B in the class. Students who attended class 80 - 90% of the time usually received a B or C in the class. Students who attended class less than 80% of the time usually received a D, F or eventually withdrew from the class.

Attendance and grades are related as shown by the results of the study. The more you attend class, the more likely you will receive a higher grade. If you improve your attendance, your grades will probably improve. Many factors affect your grade in a course. One of those factors that you have considerable control over is attendance. You can increase your opportunities for learning by attending class more often. Don't take chances with your education. Get what you pay for.

---

1. What are the variables under study?

2. Are they random variables?

3. What are the data in the study?

4. Are descriptive, inferential, or both types of statistics used?

5. What is the population under study?

6. Was a sample collected? If so, from where?

7. Where any hypotheses tested?

8. From the information given, comment on the relationship between the variables.

1-3
SAFE TRAVEL

Read the following information about the transportation industry and answer the questions.

Transportation Safety

The following shows the number of job-related injuries for each of the transportation industries for 1998.

Industry	No. of Injuries
Railroad	4520
Inter-city bus	5100
Subway	6850
Trucking	7144
Airline	9950

1. What are the variables under study?
2. Categorize each variable as quantitative or qualitative.
3. Categorize each quantitative variable as discrete or continuous.
4. Identify the level of measurement for each variable.
5. The railroad is shown as the safest transportation industry. Does that mean railroads have fewer accidents than the other industries? Explain.
6. What factors other than safety influence a person's choice of transportation industry?
7. Where any hypotheses tested?
8. From the information given, comment on the relationship between the variables.

1-4
AMERICAN CULTURE AND DRUG ABUSE

Assume you are a member of the Family Research Council and have become increasingly concerned about the drug use by professional sports players. You conduct a survey on how people believe the American culture (television, movies, magazines, & popular music) influences illegal drug use. The survey consisted of 2250 adults and adolescents from around the country. A consumer group petitions you for more information about the survey. Answer the following questions about your survey.

---

1. What type of survey did you use (phone, mail, or interview)?
2. What are the advantages and disadvantages of the surveying methods you did not use?
3. What type of scores did you use? Why?
4. Did you use a random method for deciding who would be in your sample?
5. Which of the methods (stratified, systematic, cluster, or convenience) did you use?
6. Why was that method more appropriate for this type of data collection?
7. If a convenience sample was collected, consisting of only adolescents, how would the results of the study be affected?

1-5
JUST A PINCH BETWEEN YOUR CHEEK AND GUM

As the evidence on the adverse effects of cigarette smoke grew, people tried many different ways to quit smoking. Some people tried chewing tobacco or as it was called smokeless tobacco. A small amount of tobacco was placed between the cheek and gum. Certain chemicals from the tobacco were absorbed into the bloodstream and gave the sensation of smoking cigarettes. This prompted studies on the adverse effects of smokeless tobacco. One study in particular used 40 university students as subjects. Twenty were given smokeless tobacco to chew and twenty given a substance that looked and tasted like smokeless tobacco, but did not contain any of the harmful substances. The students were randomly assigned to one of the groups. The students' blood pressure and heart rate were measured before they started chewing and twenty minutes after they had been chewing. A significant increase in heart rate occurred in the group that chewed the smokeless tobacco. Answer the following questions.

1. What type of study was this? (observational, quasi-experimental, or experimental)?
2. What are the independent and dependent variables?
3. Which was the treatment group?
4. Could the student's blood pressure be affected by knowing that they are part of a study?
5. List some possible confounding variables.
6. Do you think this is a good way to study the effect of smokeless tobacco?

1-8
MORE SMOKE THAN YOU THINK

Read the following and answer the questions below.

How Many Cigarettes Have You Smoked Today?

Cigarette smoke can't always be seen or smelled. So even if you don't smoke, you may be inhaling more smoke than you think. One cigarette emits over 4000 known chemicals and some of them are directly linked to cancer. A study was recently conducted on second hand smoke, or smoke that people inhale, while not actually smoking. Scientists collected thousands of air samples from around the country. The samples were taken in homes, offices, bars, stadiums, and other populated areas. One of the most dangerous chemicals in cigarette smoke, TFTCA, was measured. Some of the results are shown in the following table. (Equivalencies are estimates)

Location	Time (hours)	Cigarette Equivalence
Bar	2	4
Restaurant	2	1
Office	8	3
Smoker's Home	24	5
Sports Stadium	3	2
Smoker's Car	1	7

1. What was the variable under study?
2. Is it qualitative or quantitative?
3. Is it discrete or continuous?
4. Were descriptive or inferential statistics used?
5. What was the population under study?
6. Was the data collected using a random method? Explain.
7. Where there any hypotheses stated?
8. What level of measurement was used for the data?
9. Was the study observational or experimental?
10. Is the table misleading in any way?
11. What are the advantages and disadvantages of using the type of study that was used for the collection of the TFTCA samples?

2-1
WHICH GRAPH TO USE

Two of the most commonly used types of graphs are shown below. Answer the following questions about the information given.

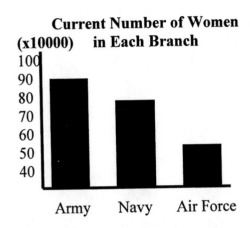
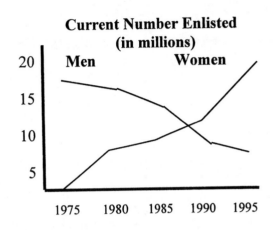

The number of women in the armed forces has been on the increase since 1975. Many of them take advantage of the college payroll plan and serve their time while investing in the future. With the rising cost of a college education, it is difficult to work part time, save enough money to go to school full time. The above charts show which armed forces attract new candidates and how the enrollment trend has been changing for men and women.

1. Do the graphs help you understand what the author is trying to get across?
2. Why were two different types of graphs used?
3. What type of variable (quantitative or qualitative) is used with the bar graph?
4. What type of variable (quantitative or qualitative) is used with the time series graph?
5. Why wasn't a graph used for college costs?
6. Does the article give you the impression that the money for college grants is unfairly distributed among the sexes?

2-2
AFTER SCHOOL ACTIVITIES

Assume you are reading the local newspaper and come across the following information. Answer the questions about the following article.

Local Kids Stats

The following lists Manatee County High School students' after-school sports activities

Students   Sport

32   basketball
56   soccer
315   football
95   band
225   baseball
185   tennis
25   wrestling
34   other

---

1. Is the information in the above article organized as a frequency distribution? If so, what type?

2. Would you consider the categories mutually exclusive?

3. Are the categories exhaustive?

4. Would it be reasonable to ask if the categories are continuous or of equal width?

5. Is it likely that more children would fall into one category as opposed to another?

6. Does equal categories in grouped frequency distributions give the impression that they are equally likely to capture data values?

7. Why do these categories have to be of equal length?

8. Assume you are in local politics and must decide how to allocate funds for high school sports activities. Use the given data to justify how the available money should be spent.

2-3
SELLING REAL ESTATE

(Computer)

Assume you are a Realtor in Bradenton, Florida. You have recently obtained a listing of the selling prices of the homes that have sold in that area in the last six months. You wish to organize that data so you will be able to provide potential buyers with useful information. Use the following data to create a histogram, frequency polygon and a cumulative frequency polygon.

142000	127000	99600	162000	89000	93000	99500
73800	135000	119500	67900	156300	104500	108650
123000	91000	205000	110000	156300	104000	133900
179000	112000	147000	321550	87900	88400	180000
159400	205300	144400	163000	96000	81000	131000
114000	119600	93000	123000	187000	96000	80000
231000	189500	177600	83400	77000	132300	166000

1. What questions could be answered more easily by looking at the histogram, rather than the listing of home prices?

2. What different questions could be answered more easily by looking at the frequency polygon rather than the listing of home prices?

3. What different questions could be answered more easily by looking at the cumulative frequency polygon rather than the listing of home prices?

4. Are there any outliers in the distribution?

5. Which graph displays outliers the best?

6. Is the distribution skewed? (Changing the number of classes can help display outliers.)

2-4
LEADING CAUSE OF DEATH

The following shows approximations of the leading causes of death among men ages 25-44 years. The rates are based on per 100,000 men. Answer the following questions about the graph.

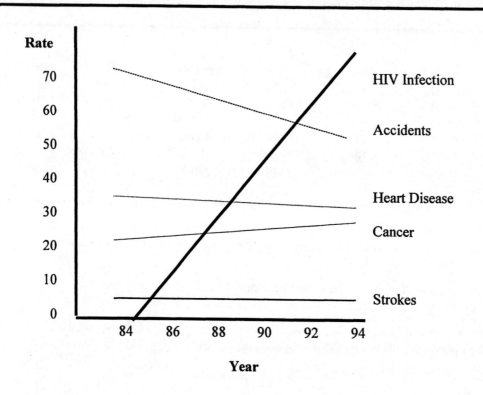

1. What are the variables in the graph?
2. Are the variables qualitative or quantitative?
3. Are the variables discrete or continuous?
4. What type of graph was used to display the data?
5. Could a Pareto chart be used to display the data?
6. Could a pie chart be used to display the data?
7. List some typical uses for the Pareto chart.
8. List some typical uses for the time series chart.

2-5
SPORTS DRINKS

Assume you get a new job as a coach for a sports team and one of your first decisions is what sports drink the team will use during practices and games. You obtain a Sports Report magazine so you can use your statistical background to help you make the best decision. The following table lists the most popular sports drinks and some important information about each of them. Answer the following questions about the table.

Drink	Calories	Sodium	Potassium	Cost
Gatorade	60	110	25	1.29
Powerade	68	77	32	1.19
All Sport	75	55	55	.89
10-K	63	55	35	.79
Exceed	69	50	44	1.59
1st Ade	58	58	25	1.09
Hydra Fuel	85	23	50	1.89

1. Is this data considered raw data?
2. Is this considered a frequency distribution?
3. Would a histogram be better than the table for certain parts of the display? If so, which parts?
4. Which column could be used for a Pareto chart?
5. Could a pie chart be used for cost per container?
6. Why might Sports Report magazine use a table instead of charts and graphs?

3-1
COLLEGE TUITION

Assume you are planning to attend college and you find the table below in a national magazine. Use the information in the table to answer the following questions.

The Cost of College in 1998

	Public	Private
Tuition and Fees	$2944	$12,033
Books and supplies	722	739
Room and board	4444	5733
Transportation	755	944
Miscellaneous	1588	1209
	$10,453	$20,658

Costs are weighted averages for the 98-99 school year. The information is for a student living on campus at a four-year college.

1. What is meant by average?
2. Assume price is a major factor in deciding where you will attend college. Is it reasonable to assume that you will be able to find a cost-competitive private college?
3. What factors influence college tuition other than it being a public or private institution?
4. Could some public colleges be more expensive than some private colleges?
5. Can one number give an accurate estimate of thousands of numbers?
6. Should you make your decision from only comparing the averages?
7. Would the report have been better if it listed a range of values or an indication of the variability in tuition costs at both public and private colleges?

3-2
TEACHER SALARIES

The following data represent salaries from a school district in Greenwood, South Carolina.

10000	11000	11000	12500	14300	17500
18000	16600	19200	21560	16400	107000

1. First, assume you work for the school board in Greenwood and do not wish to raise taxes to increase salaries. Compute the mean, median and mode and decide which one would best support your position to not raise salaries.

2. Second, assume you work for the teachers union and want a raise for the teachers. Use the best measure of central tendency to support your position.

3. Explain how outliers can be used to support one or the other position.

4. If the salaries represented every teacher in the school district, would the averages be parameters or statistics?

5. Which measure of central tendency can be misleading when a data set contains outliers?

6. When comparing the measures of central tendency, does the distribution display any skewness? Explain.

3-3
BLOOD PRESSURE

The table below lists means and standard deviations. The mean is the number before the plus/minus and the standard deviation is the number after the plus/minus. The results are from a study attempting to find average blood pressure of older adults. Use the results to answer the questions.

	Normotensive		Hypertensive	
	Men (n=1200)	Women (n=1400)	Men (n=1100)	Women (n=1300)
Age	55 ± 10	55 ± 10	60 ± 10	64 ± 10
Blood Pressure				
Systolic	123 ± 9	121 ± 11	153 ± 17	156 ± 20
Diastolic	78 ± 7	76 ± 7	91 ± 10	88 ± 10

1. Apply Chebyshev's Theorem to the systolic blood pressure of Normotensive men. At least how many of the men in the study fall within 1 standard deviation of the mean?

2. At least how many of those men in the study fall within 2 standard deviations of the mean?

Assume that blood pressure is normally distributed among older adults. Answer the following questions using the empirical rule instead of Chebyshev's Theorem.

3. Give ranges for the diastolic blood pressure (Normotensive and Hypertensive) of most older women.

4. Do the Normotensive, male, systolic blood-pressure ranges overlap with the Hypertensive, male, systolic, blood-pressure ranges?

3-4
DETERMINING DOSAGES

In an attempt to determine necessary dosages of a new drug (HDL) used to control sepsis, assume you administer varying amounts of HDL to 40 mice. You create four groups and label them low dosage, moderate dosage, large dosage and very large dosage. The dosages also vary within each group. After the mice are injected with the HDL and the sepsis bacteria, the time until the onset of sepsis is recorded. Your job as a statistician is to effectively communicate the results of the study.

---

1. Which measures of position could be used to help describe the data results?
2. If 40% of the rats in the top quartile survived after the injection, how many mice would that be?
3. What information can be given from using percentiles?
4. What information can be given from using quartiles?
5. What information can be given from using standard scores?

3-5
THE NOISY WORKPLACE

(Computer)

Assume you work for OSHA (Occupational Safety and Health Administration) and had complaints about noise levels from some of the workers at a state power plant. You charge the power plant with taking decibel readings at six different areas of the plant at different times of the day and week. The results of the data collection are listed below. Use boxplots to initially explore the data and make recommendations about which plant areas workers must be provided with protective ear wear. The safe hearing level is at approximately 120 decibels.

Area 1	Area 2	Area 3	Area 4	Area 5	Area 6
30	64	100	25	59	67
12	99	59	15	63	80
35	87	78	30	81	99
65	59	97	20	110	49
24	23	84	61	65	67
59	16	64	56	112	56
68	94	53	34	132	80
57	78	59	22	145	125
100	57	89	24	163	100
61	32	88	21	120	93
32	52	94	32	84	56
45	78	66	52	99	45
92	59	57	14	105	80
56	55	62	10	68	34
44	55	64	33	75	21

3-6
MEN'S FRAGRANCES

(Computer)

Assume you are looking for a man's fragrance and believe that you will have to spend about $20 for one of the popular brands. You decide to look up prices on the Internet to help you make a decision on which one to buy. Since price is a major concern, you decide to test your hypothesis about the average price of men's fragrances. Assume the prices you find on the Internet are randomly selected and listed below. Input the prices listed in the table and test your hypothesis.

Men's Fragrances	Price
Drakkar Noir	$15.95
Lancer	4.25
Eternity for Men	11.95
Realm for Men	27.50
Preferred Stock	9.95
Polo	12.94
Brut	3.75
Aramis	10.44
Safari for Men	13.99
Stetson	9.21
Tribute	6.95
Old Spice	3.25

1. Compute the mean, median and mode.
2. Which is the most appropriate measure of central tendency?
3. Compute the range, variance and standard deviation.
4. Which is the most appropriate measure of variability?
5. Is the distribution skewed?
6. Is the price of "Realm for Men" more that two standard deviations away from the mean? Would it be considered an outlier?
7. At what percentile is the price of Stetson?
8. Create a boxplot from the given prices.

4-1
PLAYING THE LOTTERY

Millions of people play the lottery every day. Many of those people select numbers in unusual ways in hopes of winning. Others use computers to pick the numbers or select previously chosen numbers. Of those who select their own numbers, there are several different methods used. Some use their birth dates or family member's birth dates. Other common choices are anniversary dates, shoe sizes, road signs, prices, and other environmental cues. Still others believe spirits tell them what to choose. Answer the following questions on your opinion about playing the lottery.

1. Are there any methods that you believe would increase a person's chances of winning the lottery?
2. Can an understanding of counting techniques help you improve your chances of winning the lottery?
3. What is a random number generator?
4. Will using a random number generator increase your chances of winning?
5. Do you believe the more educated a person is, the more rational they try to be about winning the lottery and typically use a random number generator to eliminate human bias?
6. How does human bias effect a person's probability of winning the lottery?
7. Does human bias effect the amount of money that is won?
8. Does being more rational increase your chances of winning?

4-2
TOSSING A COIN

Assume you are at a carnival and decide to play one of the games. You spot a table where a person is flipping a coin, and since you have an understanding of basic probability, you believe that the odds of winning are in your favor. When you get to the table, you find out that all you have to do is guess which side of the coin will be facing up after it is tossed. You are assured that the coin is fair, meaning that each of the two sides has an equally likely chance of occurring. You think back about what you learned in your statistics class about probability before you decide what to bet on. Answer the following questions about the coin tossing game.

---

1. What is the sample space?
2. What are the possible outcomes?
3. What does the classical approach to probability say about computing probabilities for this type of problem?

You decide to bet on heads, believing that it has a 50% chance of coming up. A friend of yours, who had been playing the game for a while before you got there tells you that heads has come up the last nine times in a row. You remember the law of large numbers.

4. What is the law of large numbers and does it change your thoughts about what will occur on the next toss?
5. What does the empirical approach to probability say about this problem and could you use it to solve this problem?
6. Can subjective probabilities be used to help solve this problem? Explain.
7. Assume you could win one million dollars if you could guess what the results of the next toss will be. What would you bet on? Why?

4-3
WHICH PAIN RELIEVER IS BEST?

Assume that following an injury you received from playing your favorite sport, you obtain and read information on new pain medications. In that information you read of a study that was conducted to test the side effects of two new pain medications. Use the following table to answer the questions and decide which, if any, of the two new pain medications you will use.

Number of Side Effects
12 Week Clinical Trial

Side Effect	Placebo n = 192	Drug A n = 186	Drug B n = 188
Upper Respiratory Congestion	10	32	19
Sinus Headache	11	25	32
Stomachache	2	46	12
Neurological Headache	34	55	72
Cough	22	18	31
Lower Respiratory Congestion	2	5	1

1. How many subjects were in the study?
2. How long was the study?
3. What were the variables under study?
4. What type of variables are they and what level of measurement are they on?
5. Are the numbers in the table exact figures?
6. What is the probability that a randomly-selected person was receiving a placebo?
7. What is the probability that a person was receiving a placebo or drug A? Are these mutually-exclusive events? What is the complement to this event?
8. What is the probability that a randomly-selected person was receiving a placebo or experienced a neurological headache?
9. What is the probability that a randomly-selected person was not receiving a placebo or experienced a sinus headache?

4-4
GUILTY OR INNOCENT

In July of 1964, an elderly woman was mugged in Costa Mesa, California. In the vicinity of the crime a tall, bearded man sat waiting in a yellow car. Shortly after the crime was committed, a young, tall woman, wearing her blond hair in a ponytail, was seen running from the scene of the crime and getting into the car, which sped off. The police broadcast a description of the suspected muggers. Soon afterwards, a couple fitting the description was arrested and convicted of the crime. Although the evidence in the case was largely circumstantial, the two people arrested were convicted of the crime. The prosecutor based his entire case on basic probability theory, showing the unlikeness of another couple being in that area while having all the same characteristics that the elderly woman described. The following probabilities were used.

Characteristic	Assumed Probability
Drives yellow car	1 out of 12
Man over six feet tall	1 out of 10
Man wearing tennis shoes	1 out of 4
Man with beard	1 out of 11
Woman with blond hair	1 out of 3
Woman with hair in a pony tail	1 out of 13
Woman over six feet tall	1 out of 100

1. Compute the probability of another couple being in that area with the same characteristics.

2. Would you use the addition or multiplication rule? Why?

3. Are the characteristics independent or dependent?

4. How are the computations affected by the assumptions of independence or dependence?

5. Should any court case be based solely on probabilities?

6. Would you convict the couple that was arrested even if there were no eye witnesses?

7. Comment on why in today's justice system that no person can be convicted solely on the results of probabilities?

8. In actuality, aren't most court cases based on un-calculated probabilities?

4-5A
HOW MANY WAYS COULD THERE BE?

The basis for computing probabilities involves counting. Simply put, to compute a probability, you count the number of things that you are looking for and divide that by the number of all possible things that could occur. The best way to learn about probabilities is to develop an understanding of counting techniques. Perform the following instructions and answer the following questions about counting (Try using a diagram or table to help you recognize patterns).

Assume you go to a restaurant and are going to eat one main dish and one dessert. List how many different meals are possible if the restaurant has

1. One main dish and one dessert.
2. One main dish and two desserts.
3. One main dish and three desserts.
4. Two main dishes and three desserts.
5. Three main dishes and three desserts.

Assume you are going to pair up teams for a sporting event. List how many possible pairings there are for

6. One team.
7. Two teams.
8. Three teams.
9. Four teams.
10. Five teams.
11. Twenty-five teams.

You should be able to recognize a pattern on how to continue counting multiple team pairings. This is not a very efficient way to solve counting problems other than very simple ones. Formulas can be developed from recognizing patterns and then used to solve more complicated counting problems. Try using $n(n-1)/2$ to see if it matches your answers to numbers 6-11.

4-5B
GARAGE DOOR OPENERS

Garage door openers originally had a series of four on-off switches so that home owners could personalize the frequency that opened their garage door. If all garage door openers were set at the same frequency, anyone with a garage door opener could open anyone else's garage door.

---

1. Use a tree diagram to show how many different positions four consecutive on-off switches could be in.

    After garage door openers became more popular, another set of four on-off switches was added to the systems.

2. Find a pattern of how many different positions are possible with the addition of each on-off switch.

3. How many different positions are possible with eight consecutive on-off switches?

4. Is it reasonable to assume that if you owned a garage door opener with eight switches that someone could use his/her garage door opener to open your garage door by trying all the different possible positions?

In 1989 it was reported that the ignition keys for 1988 Dodge Caravans were made from a single blank that had five cuts on it. Each cut was made at one out of five possible levels. In 1988, assume there was 420,000 Dodge Caravans sold in America.

---

5. How many different possible keys can be made from the same key blank?

6. How many different 1988 Dodge Caravans could any one key start?

    Look at the ignition key for your car and count the number of cuts on it. Assume that the cuts are made at one of any five possible levels. Most car companies use one key blank for all their makes and models of cars.

7. Conjecture how many cars your car company sold over recent years and then figure out how many other cars your car key could start. What would you do to decrease the odds of someone being able to open another vehicle with their key?

4-5C
SEATING ARRANGEMENTS

Not all seemingly-easy problems can be solved by mathematicians, even if they use computers. Problems involving the use of permutations can become very difficult even with small numbers. Remember that when counting the number of ways to order things you can use factorial notation. Let's look at some examples.

You are planning a family dinner for you and some of your relatives. Remembering that they do not all get along, you decide to list different ways they can sit at the dinner table. Answer the following questions.

---

1. How many ways could 10 people sit in ten chairs?

2. If you started listing a different way every second continually, would you be able to write out the complete list by next month?

   Many practical problems require the use of factorial notation to solve them. A famous problem in the late 1970's was called the traveling salesman problem. Scientists tried to find the shortest possible route through 50 cities.

3. How many ways could you arrange 50 cities? How about 100 cities?

4. Does your calculator even compute 100 factorial? Try 500 factorial.
   One hundred factorial is about 10 to the 200th power, or 1 with 200 zeros after it. How long do you think it would take a computer to solve it? If every electron in the universe were a computer that could do one billion computations per second, it would take all those computers 100,000,000,000 years to complete the calculation.

5. List five realistic applications of where a business could run more efficiently if it could find shortest possible routes.

4-6
WHICH METHOD TO USE

One of the biggest problems for students is deciding on what method or formula to use to solve a probability problem. You first need to identify how many events are occurring. If it is one simple event, you set up the classical ratio. If it is one compound event, you use the addition rule. If it is more than one event, you will be multiplying something. If you have a large number of events, your multiplications can be aided by a formula. Students typically fail to identify type of problem when dealing with multiple events that are either independent or dependent. Use the following activity to develop a better understanding of which probability methods to use.

Assume you are going to guess at 5 multiple-choice questions. There are 5 possible answers for each questions (A, B, C, D, or E).

1. How many events are there?

2. Are they independent or dependent?

3. What are the odds of getting them all correct?

4. What are the odds of someone guessing A for every problem?

Assume the 5 questions were matching and you could only use each answer one time.

5. How many events are there?

6. Are they independent or dependent?

7. What are the odds of getting them all correct?

8. What is the main difference between the two problems?

4-7A
SUBTLE DIFFERENCES

Think about how you could solve the following problem?
Problem: How many different ways could you answer 10 true-false questions?
Answer the following questions.

---

1. Could you use a tree diagram to solve this problem?

2. Could you use the multiplication rule to solve this problem?

3. Since it doesn't matter in what order you answer the questions, could you use the combination formula to solve the problem?

4. If order matters, could you use the permutation formula?

5. What would you use if the ten problems were matching?

6. How can you tell when to use a tree diagram, multiplication rule, permutations or combinations?

4-7B
MISLEADING WORDS

Assume you read an article in a consumer activist magazine that states 60,000 people a day are affected by dig-ups (construction projects that accidentally cut communication lines) and are denied access to the public switched network for an average of two to three hours. The article states that many people are left at risk when an emergency occurs and they cannot communicate their need for help. What is the probability of being effected by a dig-up?

1. Determine the sample space.
2. Is the addition rule needed for any part of the problem?
3. How many events are there?
4. Are the events independent or dependent?
5. Are rare probabilities such as this one, or getting hit by a meteor, useful in trying to predict when things will occur?
6. Are there other things that influence where dig-ups occur?
7. Could you use a mathematical formula?
8. How would you compute the probability of you being effected by a dig-up?

5-2A
DRINKING AND DRIVING

The following chart shows approximations of how drinking alcohol can increase the probability of being involved in an automobile accident. Use the chart to answer the questions below.

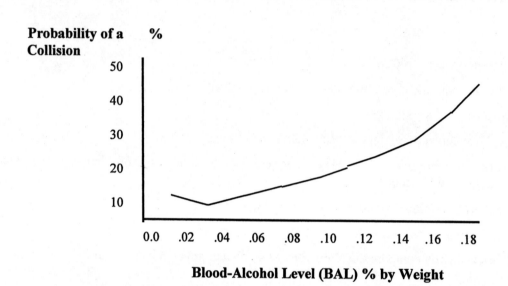

1. What are the variables under study?
2. Are they discrete or continuous variables?
3. Would you consider this a probability distribution?
4. Would it be considered a theoretical probability distribution?
5. What is the probability of a collision if the BAL is 0.10?
6. How many different BALs are there?
7. Does each of the BALs have it's own probability?
8. Could a discrete probability distribution be a good estimate of this distribution?

5-2B
DROPPING COLLEGE COURSES

Use the following table to answer the questions.

Reasons for Dropping a College Course

Reason	Frequency	Percentage
Too Difficult	45	
Illness	40	
Change in Work Schedule	20	
Change Major	14	
Family Related Problems	9	
Money	7	
Miscellaneous	6	
No Meaningful Reason	3	

1. What is the variable under study? Is it a random variable?
2. How many people were in the study?
3. Complete the table.
4. From the information given, what is the probability that a student will drop a class because of; Illness? Money? Changing their major?
5. Would you consider the information in the table to be a probability distribution?
6. Are the categories mutually exclusive?
7. Are the categories independent?
8. Are the categories exhaustive?
9. Are the two requirements for a discrete probability distribution met?

5-3
EXPECTED VALUE

On March 28, 1979, the nuclear generating facility at Three Mile Island, Pennsylvania, began discharging radiation into the atmosphere. People exposed to even low levels of radiation can experience health problems ranging from very mild to severe, even causing death. A local newspaper reported that 11 babies were born with kidney problems in the three county area surrounding the Three Mile Island nuclear power plant. The expected value for that problem in infants in that area was 3. Answer the following questions.

---

1. What does expected value mean?

2. Would you expect the exact value of 3 all the time?

3. If a news reporter stated that the number of cases of kidney problems in newborns was nearly four times as much as was usually expected, do you think pregnant mothers living in that area would be overly concerned?

4. Is it unlikely that 11 occurred by chance?

5. Are there any other statistics that could better inform the public?

6. Assume that 3 out of 2500 babies were born with kidney problems in that three county area the year before the accident. Also assume that 11 out of 2500 babies were born with kidney problems in that three county area the year after the accident. What is the real percent of increase in that abnormality?

7. Do you think that pregnant mothers living in that area should be overly concerned after looking at the results in terms of rates?

5-4
UNSANITARY RESTAURANTS

Health officials routinely check sanitary conditions of restaurants. Assume you visit a popular tourist spot and read in the newspaper that in 3 out of every 7 restaurants checked, there were unsatisfactory health conditions found. Assuming you are planning to eat out 10 times while you are there on vacation, answer the following questions.

1. How likely is it that you will eat at three restaurants with unsanitary conditions?
2. How likely is it that you will eat at four or five restaurants with unsanitary conditions?
3. Explain how you would compute the probability of eating in at least one restaurant with unsanitary conditions. Could you use the complement to solve this problem?
4. What is the most likely number to occur in this experiment?
5. How variable will the data be around the most likely number?
6. Is this a binomial distribution?
7. If it is a binomial distribution, does that mean that the likelihood of a success is always 50% since there are only two possible outcomes?

Check your answers using the following computer-generated table.

Mean = 4.3   St. Dev. = 1.56557

X	P(x)	Cum Prob
0	.00362	.00362
1	.02731	.03093
2	.09272	.12365
3	.18651	.31016
4	.24623	.55639
5	.22291	.77930
6	.14013	.91943
7	.06041	.97983
8	.01709	.99692
9	.00286	.99979
10	.00022	1.00000

5-5
HOW SAFE ARE YOU?

(Computer)

Assume one of your favorite activities is mountain climbing. When you go mountain climbing you have several safety devices to keep you from falling. You notice that attached to one of your safety hooks is a reliability rating of 97%. You estimate that throughout the next year you will be using this device about 100 times. Answer the following questions.

---

1. Does that mean there is a 97% chance that the device will not fail any of the 100 times?

2. What is the probability of at least one failure?

3. What is the compliment of this event?

4. Is this a binomial experiment?

5. Can you use the binomial probability formula? Why or Why not?

6. How much safer would it be to use a second safety hook independently of the first?

7. Can the Poisson distribution be used to approximate the binomial distribution?

8. Use the Poisson distribution to find the probability of exactly 5 failures out of 100 times using the safety hook.

9. When is the Poisson distribution a good approximation to the binomial?

5-6
CHOOSING THE BEST DISTRIBUTION

The following shows the probability of buying a defective light bulb. Use the graph to help you answer the questions. Zero represents a non-defective light bulb. One represents a defective light bulb.

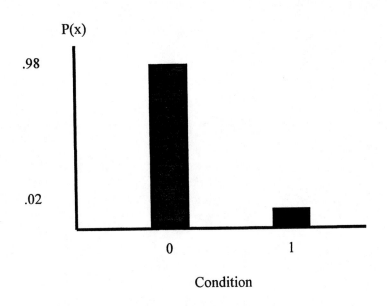

1. Is this a random variable?
2. Is this a probability distribution?
3. What is the expected value?
4. What is the probability of 4 successes out of 5 trials? Use the binomial formula and the Poisson distribution formula. Compare your results.
5. Assume fifty of the light bulbs are in a large box. What is the probability of selecting 20 of those light bulbs and getting 1 defect?
6. What is the difference between the binomial and hypergeometric distributions?

6-1
AVERAGE INCOME

(Computer)

1. Input the average per-capita incomes that are listed in the table and compute measures of central tendency.

2. Create a histogram using an appropriate number of classes. From looking at the histogram, is the per capita income approximately normally distributed?

3. Check the measures of central tendency and comment on how they indicate the type of distribution (no skewness, positively skewed, or negatively skewed).

4. Is it possible that a positively-skewed distribution could have the mode less than the mean less than the median?

5. Create a set of data values where the mean is greater than the median and both are greater than the mode, but the distribution is negatively skewed. Does this contradict the relationships stated in the textbook? Explain.

Dist of Columbia	$33,443	Wisconsin	$21,819
Connecticut	31,209	Kansas	21,712
New York	29,488	Oregon	21,644
Maryland	28,048	Ohio	21,331
New Jersey	27,255	Nebraska	20,888
Massachusetts	26,880	Indiana	20,777
Delaware	26,045	Vermont	20,323
Illinois	26,021	North Carolina	19,898
New Hampshire	25,992	South Carolina	19,888
Nevada	25,474	South Dakota	19,576
California	25,115	Wyoming	19,432
Hawaii	24,703	Iowa	19,210
Alaska	24,353	Arizona	18,607
Virginia	24,004	Alabama	18,342
Minnesota	23,990	Maine	17,903
Washington	23,843	Utah	17,304
Rhode Island	23,822	Montana	17,222
Michigan	23,744	Idaho	16,443
Colorado	23,402	Louisiana	16,321
Pennsylvania	23,230	Kentucky	16,290
Florida	23,199	North Dakota	16,288
Tennessee	22,789	Oklahoma	16,231
Texas	22,196	West Virginia	16,222
Georgia	22,103	New Mexico	16,195
Missouri	21,954	Arkansas	15,333
		Mississippi	14,021

6-2
NORMAL HEALTH

The following table lists base-line characteristics of two randomly-selected groups of men and women. The exercise group consisted of individuals who participated in vigorous exercise for at least 20 minutes a day, 3 times a week, for at least the last 6 months. The no exercise group consisted of individuals who had not engaged in any type of vigorous exercise over the last 6 months.

Characteristic	Exercise (n=23)	No Exercise (n=23)
Age	57 ± 10	58 ± 11
Weight	127 ± 16	125 ± 15
Height	67 ± 5	68 ± 8
Heart Rate	73 ± 9	74 ± 9
Systolic Blood Pressure	137 ± 10	139 ± 15
Diastolic Blood Pressure	88 ± 7	88 ± 6
Body Mass	33 ± 5.8	31 ± 4.4

The numbers shown are means and standard deviations.
The body mass index is the weight in kilograms divided by the square of the height in meters.

Assume base-line characteristics are normally distributed.

1. How many subjects' heart rates fell within one standard deviation of the mean?

2. How many subjects' diastolic blood pressure from the no-exercise group fell between 82 and 94?

3. What percentage of the subjects' weight from the exercise group fell between 65 and 129?

4. Does there seem to be any significant differences between the exercise group and the no-exercise group on any of the base-line characteristics?

6-3
THE Z SCORE

(Computer)

1. Select or create a quantitative data file.
2. Compute descriptive statistics and create a histogram.
3. Subtract the mean from every data value in the data set and compute the new descriptive statistics.
4. Create a histogram of the new data set using the same number of classes as the first histogram and compare histograms and descriptive statistics.
5. Divide every data value by the original sample standard deviation and compute the new descriptive statistics.
6. Create a histogram of the new data set using the same number of classes as the first two histograms and compare histograms and descriptive statistics.
7. Summarize your findings.
8. Are the original data values approximately normally distributed?
9. Is the last set of data values a standard normal distribution? Why or why not?
10. What would have resulted if the population standard deviation instead of the sample standard deviation would have been used when dividing into every data value?
11. Does standardizing scores reduce the variability of the original data set?

6-4
SMART PEOPLE

Assume you are thinking about starting a MENSA chapter in your home town of Visiala, California which has a population of about 10,000 people. You need to know how many people would qualify for MENSA, which requires an IQ of at least 130. You realize that IQ is normally distributed with a mean of 100 and a standard deviation of 15. Complete the following.

---

1. Find the approximate number of people in Visiala that are eligible for MENSA.

2. Is it reasonable to continue your quest for a MENSA chapter in Visiala?

3. How would you proceed to find out how many of the eligible people would actually join the new chapter? Be specific about your methods of gathering data.

4. What would be the minimum IQ score needed if you wanted to start an Ultra-Mensa club that included only the top 1% of IQ scores.

6-5
CENTRAL LIMIT THEOREM

(Computer)

Twenty students from a statistics class each collected a random sample of times on how long it took students to get to class from their home. All of the sample sizes were 30. The resulting means are listed below.

Student	Mean	SD	Student	Mean	SD
1	22	3.7	11	27	1.4
2	31	4.6	12	24	2.2
3	18	2.4	13	14	3.1
4	27	1.9	14	29	2.4
5	20	3.0	15	37	2.8
6	17	2.8	16	23	2.7
7	26	1.9	17	26	1.8
8	34	4.2	18	21	2.0
9	23	2.6	19	30	2.2
10	29	2.1	20	29	2.8

1. The students noticed that everyone had different answers. If you randomly sample over and over again from any population, with the same sample size, will the results ever be the same?

2. The students wondered whose results were right. How can they find out what the population mean and standard deviation are?

3. Input the means into the computer and check to see if the distribution is normal.

4. Check the mean and standard deviation of the means. How do these values compare to the students' individual scores?

5. Is the distribution of the means a sampling distribution?

6. Check the sampling error for students number 3, 7 and 14.

7. Compare the standard deviation of the sample of the 20 means. Is that equal to the standard deviation from student number 3 divided by the square of the sample size? How about for student number 7, or 14?

6-6
HOW SAFE ARE YOU?

(Computer)

Assume one of your favorite activities is mountain climbing. When you go mountain climbing you have several safety devices to keep you from falling. You notice that attached to one of your safety hooks is a reliability rating of 97%. You estimate that throughout the next year you will be using this device about 100 times. Answer the following questions.

---

1. Does a reliability rating of 97% mean that there is a 97% chance that the device will not fail any of the 100 times?

2. What is the probability of at least one failure?

3. What is the compliment of this event?

4. Can this be considered a binomial experiment?

5. Can you use the binomial probability formula? Why or Why not?

6. Find the probability of at least two failures.

7. Can you use the normal distribution to accurately approximate the binomial distribution? Explain why or why not?

8. Is correction for continuity needed?

9. Use the technology to verify your answers?

10. How much safer would it be to use a second safety hook independently of the first?

6-7
THE PROBLEM WITH NORMAL HEALTH

Many people associate the norm as being okay or in a desirable position. Likewise, they also associate abnormal with not being okay or an undesirable position. One example is blood pressure. The so-called normal systolic blood pressure was determined about 20 years ago from the results of a study done by the National Institute of Health. The results came from about 2000 men and women between the ages of 15 to 20. Many people have used that as a guideline for measuring their own health. More recently it has been found that many other variables influence blood pressure such as gender, race, age and heredity. For example, the normal systolic blood pressure for a 50 year-old male is about 150. Answer the following questions.

---

1. Does that mean it is okay or healthy for a 50 year-old man to have a systolic blood pressure of 150?

2. Another example from the health field is blood serum cholesterol. The norm for middle-aged Americans is around 200. Would you feel comfortable about your health if yours was that high? More and more research is showing the link between heart disease and elevated cholesterol levels.

3. Another example of a misleading norm is the results of a hard exam. You may be right in the middle of a normally distributed population of exam scores where everybody failed the exam. List five more examples where the norm is not a desirable place to be.

4. List five examples when you would want to be above the norm.

5. List five examples when you would want to be below the norm.

6. How could researchers use the idea of being well above the mean or well below the mean in a normally distributed population as a desirable result from an experiment or test?

7-2A
CONTRACTING INFLUENZA

Use the following table to answer the questions.

Percentage of people who reported contracting influenza by sex and race/ethnicity.

Influenza

Characteristic	Percent	(95% CI)
Sex		
Men	48.8	(47.1% - 50.5%)
Women	51.5	(50.2% - 52.8%)
Race/Ethnicity		
White	52.2	(51.1% - 53.3%)
Black	33.1	(29.5% - 36.7%)
Hispanic	47.6	(40.9% - 54.3%)
Other	39.7	(30.8% - 48.5%)
Total	50.4	(49.3% - 51.5%)

Forty-nine states and the District of Columbia participated in the study. Weighted means were used. The sample size was 19,774. There were 12,774 women and 7000 men.

1. Explain what (95% CI) means?
2. How large is the error for men reporting influenza?
3. What is the sample size?
4. How does sample size effect the size of the confidence interval?
5. Would the confidence intervals be larger or smaller for a 90% CI using the same data?
6. Where does the 51.5% under influenza for women fit into its associated 95% CI?

7-2B
MAKING DECISIONS WITH CONFIDENCE INTERVALS

Assume you work for Kimberly Clark Corporation, the makers of Kleenex. The job you are presently working on requires you to decide how many Kleenexes are to be put in the new automobile glove compartment boxes. Complete the following.

1. How will you decide on a reasonable number of Kleenexes to put in the boxes?

2. When do people usually need Kleenexes?

3. What type of data collection technique would you use?

4. Assume you found out that from your sample of 85 people, on average about 57 Kleenexes are used throughout the duration of a cold, with a standard deviation of 15. Use a confidence interval to help you decide how many Kleenexes will go in the boxes.

5. Explain how you decided on how many Kleenexes will go in the boxes.

7-3
SPORT DRINK DECISION

Assume you get a new job as a coach for a sports team and one of your first decisions is deciding on which sports drink the team will use during practices and games. You obtain a *Sports Report* magazine so you can use your statistical background to help you make the best decision. The following table lists the most popular sports drinks and some important information about each of them. Answer the following questions about the table.

Drink	Calories	Sodium	Potasium	Cost
Gatorade	60	110	25	1.29
Powerade	68	77	32	1.19
All Sport	75	55	55	.89
10-K	63	55	35	.79
Exceed	69	50	44	1.59
1st Ade	58	58	25	1.09
Hydra Fuel	85	23	50	1.89

1. Would this be considered a small sample?
2. Compute the mean cost per container and create a 90% confidence interval about that mean. Do all of the costs per container fall inside the confidence interval? If not, which ones do not?
3. Are there any you would consider outliers?
4. How many degrees of freedom are there?
5. If cost is a major factor influencing your decision would you consider cost per container or cost per serving?
6. List which drink you would recommend and why.

7-4
DECEIVING STANDARD ERRORS

(Computer)

Answer the questions using the following table.

How Americans Voted	Clinton	Dole
All voters	48.5	41.3
Males	44.2	44.3
Females	55.2	39.1
Whites	42.8	45.9
Blacks	88.3	11.4
College Graduates	42.9	44.2
Postgraduate	55.6	40.1
> 65	59.4	41.3
< 25	67.1	32.5

The data are percentages from the College Review News Service. Exit polls of 16,883. Margin of Error of ± 1%.

---

1. How many people were in the survey?
2. Create a 99% confidence interval about the percentage of females who voted for Clinton.
3. How does the confidence interval relate to the margin of error of ± 1% ?
4. How did the study come up with the margin of error of 1%?
5. What would the margin of error be if the sample size was 100,000?
6. What would the margin of error be if the sample size was 100,000 from around the world which has a population size of over 4 billion?
7. Comment on sample size and the standard error of the proportion.

7-5
FIGHTING DEPRESSION

The table below shows the results of a study conducted to test the effectiveness of Prozac on depression. Patients were tested at the beginning of the experiment, after low dosage treatments, and then again after an increase in dosage of Prozac. Use the table to answer the questions below.

**Risk Ratios for Side Effects After Adjustments
For Base-Line Characteristics**

Variable	Parameter Estimate	Variances (95% CI)	Chi Square	P Value
Depression at entry	1.53	4.63 (2.02-10.59)	13.15	<0.001
Age > 25	.85	2.34 (1.11- 4.96)	4.96	0.026
Prozac	0.03	1.03 (1.00- 1.05)	4.73	0.030
Pain after six hours	1.55	4.71 (1.62-13.71)	8.05	0.005
Age > 45	0.58	1.79 (0.68- 4.75)	1.37	0.240
Prozac increase of 1 mg/ml	0.03	1.03 (1.01- 1.06)	5.47	0.051

n = 1407

1. How are the Parameter Estimates computed?
2. How are the 95% CI's for Variances computed?
3. Which Chi-Square values are significant at the 1% significance level?
4. Which Chi-Square values are significant at the 5% significance level?
5. Which significance level do you think they used? Why?
6. How many people were in the study?
7. What do they mean by the ***Parameter Estimate***?
8. What is the Chi-Square formula used for?

7-6
HAZARDS OF SMOKING

(Computer)

Use the table below to answer the following questions.

1. Find the projected average number of smokers under the age of 18 in the United States.
2. Find the standard deviation of the given data and estimate the margin of error.
3. Create a confidence interval about the proportion of young people in Florida who will become regular smokers and who will eventually die of smoking related illnesses (use n = 150,000)
4. Estimate the sample size needed if you want to cut the error in half.

The following table lists the projected number of smokers under age 18 in each state that will eventually die from smoking related illnesses.

State	Number	State	Number
Dist of Columbia	133,443	Wisconsin	121,819
Connecticut	121,209	Kansas	51,712
New York	114,488	Oregon	21,644
Maryland	148,048	Ohio	141,331
New Jersey	127,255	Nebraska	80,888
Massachusetts	76,880	Indiana	80,777
Delaware	46,045	Vermont	30,323
Illinois	56,021	North Carolina	119,898
New Hampshire	95,992	South Carolina	99,888
Nevada	65,474	South Dakota	69,576
California	85,115	Wyoming	39,432
Hawaii	34,703	Iowa	79,210
Alaska	74,353	Arizona	48,607
Virginia	54,004	Alabama	78,342
Minnesota	63,990	Maine	117,903
Washington	43,843	Utah	117,304
Rhode Island	123,822	Montana	217,222
Michigan	73,744	Idaho	116,443
Colorado	63,402	Louisiana	86,321
Pennsylvania	33,230	Kentucky	66,290
Florida	43,199	North Dakota	46,288
Tennessee	82,789	Oklahoma	116,231
Texas	62,196	West Virginia	116,222
Georgia	42,103	New Mexico	66,195
Missouri	121,954	Arkansas	75,333
Mississippi	94,021		

8-1
EGGS AND YOUR HEALTH

Answer the questions about the following information.

The Incredible Edible Egg company recently found that eating eggs does not increase a person's blood serum cholesterol. Five hundred subjects participated in a study that lasted for two years. The participants were randomly assigned to either a no-egg group or a moderate-egg group. The blood serum cholesterol levels were checked at the beginning and at the end of the study. Overall, the groups' levels were not significantly different. The company reminds us that eating eggs is healthy if done in moderation. Many of the previous studies relating eggs and high blood serum cholesterol jumped to improper conclusions.

---

1. What prompted the study?
2. What is the population under study?
3. Was a sample collected?
4. What was the hypothesis?
5. Was data collected?
6. Were any statistical tests run?
7. What was the conclusion?

8-2
QUITTING SMOKING

Assume you are part of a research team that compares products designed to help people quit smoking. The Condor Consumer Products Company would like more specific details about the study to be made available to the scientific community. Review the following and answer the questions below about how you would have conducted the study.

New StopSmoke

No method has been proven more effective. StopSmoke provides significant advantages to all other methods. StopSmoke is simpler to use and it requires no weaning. StopSmoke is also significantly less expensive than the leading brands. StopSmoke's superiority has been proven in two independent studies.

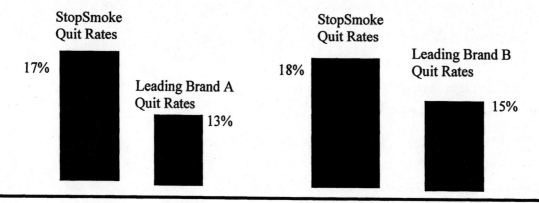

---

1. What were the statistical hypotheses?
2. What were the null hypotheses?
3. What were the alternative hypotheses?
4. Where any statistical tests run?
5. Were one- or two-tailed tests run?
6. What were the levels of significance?
7. If a type I error was committed, explain what it would have been?
8. If a type II error was committed, explain what it would have been?
9. What did the studies prove?
10. Two statements are made about significance. The one states that StopSmoke provides significant advantages and the other states that StopSmoke is significantly less expensive than other leading brands. Are they referring to statistical significance? What other type of significance is there?

8-3
BEE STING THERAPY

Use the table to answer the following questions.

Results of an Analysis of Side Effects Among Patients Using Bee Sting Therapy
n = 1402

Characteristic	Relative Risk	95% CI	P-value
Condition at base-line	2.90	2.10-3.90	<0.001
Age	1.02	1.01-1.03	0.01
Prior therapy	1.40	1.20-1.80	0.001
After therapy	0.65	0.53-0.83	0.04
Male	1.10	0.87-1.40	0.42
White	1.06	0.83-1.35	0.16
Injection	0.98	0.79-1.21	0.02

1. How many patients were in the study?
2. Is this considered a large sample?
3. What does CI mean?
4. Which characteristics were significant at the 0.01 level?
5. Are the numbers in the Relative Risk column the test statistics?
6. How do the Relative Risk numbers relate to those in the 95% CI column?
7. How does falling outside a confidence interval relate to the critical value used for rejection of the null hypothesis?
8. What does the *P*-value tell you?
9. How is the *P*-value found?

8-4
MEN'S FRAGRANCES

(Computer)

You are looking for a man's fragrance and believe that you will have to spend about $20 for one of the popular brands. You decide to look up prices on the Internet to help you make a decision on which one to buy. Since price is a major concern, you decide to test your hypothesis about the average price of men's fragrances. Assume the prices you find on the Internet are randomly selected and listed below. Input the prices listed in the table and test your hypothesis.

Men's Fragrances	Price
Drakkar Noir	$15.95
Lancer	4.25
Eternity for Men	11.95
Realm for Men	27.50
Preferred Stock	9.95
Polo	12.94
Brut	3.75
Aramis	10.44
Safari for Men	13.99
Stetson	9.21
Tribute	6.95
Old Spice	3.25

1. What is your hypothesis?
2. Is this considered a small sample when testing a hypothesis about a population mean?
3. What assumptions must be met for the hypothesis test?
4. Which probability distribution will you use?
5. How many degrees of freedom are there?
6. Would you run a one- or two-tailed test? Why?
7. Is there enough support from the results of the hypothesis test for you to reject your hypothesis?
8. What range of prices do most men's fragrances fall within?

8-5
HAIR CONTROL

Read the following advertisement and answer the questions.

Effective Options in Hair Control

All of the hair products listed below are different, but they all have a lot in common. They're all made by Hairco, the leader in men's hair control. They're also all very effective. In fact, hair controls like these are 97% effective when used correctly and consistently. That's because they all contain Nicimide-9, the number one doctor recommended tonic. They are also easy to use and safe.

Failure Rate (one year).

Type	Correct and Consistent Use	Typical Use
HairMec (foam)	11%	18%
HairTrol (gel)	10%	14%
HairBack (liquid)	17%	28%
HairCol (paste)	9%	15%

1. Is the advertisement misleading in any way?
2. How do you think the data was collected?
3. Was a sample size listed?
4. What type of measurement error could have existed?
5. Is there any indication of how accurate the percentages are?
6. How may the results be biased since they used the results of over 1 year of use?
7. How can hypothesis testing about a sample proportion be used to create a more systematic study?

8-6
TESTING GAS MILEAGE CLAIMS

(Computer)

Assume that you are working for the Consumer Protection Agency and have recently been getting complaints about the highway gas mileage of the new Dodge Caravans. Chrysler Corporation agrees to allow you to randomly select 40 of their new Dodge Caravans to test the highway mileage. Chrysler claims that their caravans get 28 MPG on the highway. Your results show a mean of 26.7 and a standard deviation of 4.2. You support Chryslers' claim.

---

1. Show why you support Chryslers' claim by listing the p-value from your output. After more complaints, you decide to test the variability of the mpg on the highway. From further questioning of Chryslers' quality control engineers, you find they are claiming a standard deviation of 2.1.

2. Test the claim about the standard deviation.

3. Write a short summary of your results and any necessary action that Chrysler must take to remedy customer complaints.

4. State your position about the necessity to perform tests of variability along with tests of the means.

8-7
CONFIDENCE INTERVALS AND HYPOTHESIS TESTING

(Computer)

Hypothesis testing and testing claims with confidence intervals are two different approaches that lead to the same conclusion. In the following activities, you will compare and contrast those two approaches.

Assume you are working for the Consumer Protection Agency and have recently been getting complaints about the highway gas mileage of the new Dodge Caravans. Chrysler Corporation agrees to allow you to randomly select 40 of their new Dodge Caravans to test the highway mileage. Chrysler claims that their vans get 28 mpg on the highway. Your results show a mean of 26.7 and a standard deviation of 4.2. You are not certain if you should create a confidence interval or run a hypothesis test. You decide to do both at the same time.

---

1. Draw a normal curve labeling the critical values, critical regions, test statistic, and the population mean. List the significance level, the null and alternative hypothesis.

2. Draw a confidence interval directly below the normal distribution labeling the sample mean, the error, and the boundary values.

3. Explain which parts from each approach are the same and which parts are different.

4. Draw a picture of a normal curve and confidence interval where the sample and hypothesized means are equal.

5. Draw a picture of a normal curve and confidence interval where the lower boundary of the confidence interval is equal to the hypothesized mean.

6. Draw a picture of a normal curve and confidence interval where the sample mean falls in the left critical region of the normal curve.

8-8
HOW MUCH NICOTINE IS IN THOSE CIGARETTES?

A tobacco company claims that its best selling cigarettes contain at most 40mg of nicotine. This claim is tested at the 1% significance level by using the results of 15 randomly selected cigarettes. The mean was 42.6 mg and the standard deviation was 3.7 mg. Evidence suggests that the distribution of nicotine is normally distributed. Information from a computer output of the hypothesis test is listed below.

$$
\begin{aligned}
&\text{Sample mean} = 42.6 \\
&\text{Sample st. dev} = 3.7 \\
&\text{Sample size} = 15 \\
&\text{Degrees of freedom} = 14 \\
&\text{P-value} = .992 \\
&\text{Significance level} = .01 \\
&\text{Test statistic} \quad t = 2.72155 \\
&\text{Critical value} \quad t = 2.62610
\end{aligned}
$$

---

1. What are the degrees of freedom?

2. Is this a large or small sample test?

3. Is this a comparison of one or two samples?

4. Is this a right, left, or two-tailed test?

5. From observing the P-value, what would you conclude?

6. By comparing the test statistic to the critical value, what would you conclude?

7. Is there a conflict in this output? Explain.

8. What has been proven in this study?

9-1
DOES REPLICATION CONFIRM?

Read the following report and answer the questions.

Does Replication Confirm?

In two randomized, double-blind trials, researchers studied how often patients experienced tremors during one year of daily injections of 30 mg of a drug or placebo. In the first study, patients who took the drug (experimental group) had a mean of 1.2 tremors compared with 3.5 for those who used the placebo (control group). FDA officials questioned the statistical accuracy of the results because it was unclear if the patients kept accurate records. The second study was a replication of the first. Two hundred patients who took the drug had a mean of 3.11 tremors as compared to 5.89 who took the placebo (n=155).

1. How many groups were in each study?
2. Are the groups independent or dependent?
3. In the second study how many people were in each group?
4. What type of data was collected?
5. Were control and experimental groups used?
6. What were the results of the original study?
7. What was the concern in the original study?
8. Did the second study results support the first?

9-2
ARTIFICIAL SWEETENERS AND YOUR HEALTH

(Computer)

The chart below shows selected characteristics of patients from a study conducted on the effects of artificial sweeteners. High use indicates the patient averaged eight or more servings of artificial sweetener per day. Low use indicates the patient averaged two or less servings per day.

Selected Characteristics of Patients in the Study

Characteristic	All Patients (n = 1397)	High Use (n=569)	Low Use (n=828)	P-value
Mean Age	58.4	59.1	57.2	.17
Male (%)	72	66	77	.16
White (%)	34	33	35	.70
Smoke (%)	37	36	33	.03
Hypertension (%)	55	48	66	.01
Heart Attack	29	30	27	.001
Pain Episodes 1st day	4.4	2.2	6.0	.09
Duration of Pain (hr)	1.7	.98	2.2	.005

1. How many subjects were in the study?
2. How were the groups separated?
3. How large was each group?
4. Are there any significant differences at the 0.05 level of significance between the groups for the listed characteristics? At the 0.01 level?

Input the following into a computer. For group 1; n = 569, with a mean and standard deviation of 30 and 8 respectively. For group 2; n = 828, with a mean and standard deviation of 27 and 11 respectively.

5. Test for any significant differences between the group means and summarize your results.

9-3
VARIABILITY AND AUTOMATIC TRANSMISSIONS

(Computer)

Assume the following data values are from the June 1996 issue of *Automotive Magazine*. An article compared various parameters of American and Japanese made sports/sporty cars. This report centers on the price of an optional automatic transmission. Which country has the most variability in the price of automatic transmissions? Input the data and answer the following questions.

Japanese Cars		American Cars	
Nissan 300ZX	$1940	Dodge Stealth	$2363
Mazda RX7	$1810	Saturn	$1230
Mazda MX6	$1871	Mercury Cougar	$1332
Nissan NX	$1822	Ford Probe	$ 932
Mazda Miata	$1920	Eagle Talon	$1790
Honda Prelude	$1730	Chevy Lumina	$1833

1. What is the null hypothesis?
2. What test statistic is used to test for any significant differences in the variances?
3. Is there a significant difference in the variability in the prices between the two car companies?
4. What effect does a small sample size have on the standard deviations?
5. What degrees of freedom are used for the statistical test?
6. Could two sets of data have significantly different variances without having significantly different means?
7. How might significantly different variances effect the results of the two-sample, small independent, t test?

9-4
TOO LONG ON THE TELEPHONE

A company collects data on the lengths of telephone calls made by employees in two different divisions. The mean and standard deviation for the sales division are 10.26 and 8.56 respectively. The mean and standard deviation for the shipping and receiving division are 6.93 and 4.93 respectively. A hypothesis test was run and the computer output follows.

Test statistic .... F = 3.07849
P-value .... = .01071
Significance level .... = .01

Degrees of freedom .... df = 56
Conf. int. limits .... = −.18979, 6.84979
Test statistic .... t = 1.89566
Critical value.................... t = −2.0037, 2.0037
P-value............................ = .06317
Significance level .... = .05

---

1. Are the samples independent or dependent?
2. Why was the F test run?
3. Were the results of the F test significant?
4. How many were in the study?
5. Which number from the output is compared to the significance level to check if the null hypothesis should be rejected?
6. Which number from the output gives the probability of a type-I error that is calculated from the sample data?
7. Which number from the output is the result of dividing the two sample variances?
8. Was a right, left, or two-tailed test run? Why?
9. What are your conclusions?
10. What would your conclusions be if the level of significance was initially set at 0.10?

9-5
MEMORIZING AND FAMILIARITY

(Computer)

Assume the follow data were collected from a study conducted on the effect of context on memorizing numbers. It was believed that imbedding numbers in sentences that students were familiar with would help them remember the numbers. The Part 1 column represents the number of numbers memorized without being imbedded in sentences. The Part 2 column represents the number of numbers memorized that were imbedded into familiar sentences.

Subject	Part 1	Part 2
1	2	3
2	6	12
3	5	3
4	5	2
5	0	5
6	5	5
7	5	8
8	2	6
9	3	6
10	2	7
11	4	4
12	3	5
13	7	5
14	3	4
15	9	5
16	3	2
17	6	5
18	3	4
19	9	4
20	1	4

1. What is the purpose of the study?
2. Are the groups independent or dependent?
3. What is the research question?

Input the data into a computer or calculator and answer the following questions.

4. How many degrees of freedom are there?
5. What is the null hypothesis?
6. Is there a significant difference between the pretest and posttest scores?
7. Could an independent test have been used to test the same hypothesis?

9-6
JOBS AND PARENTING

(Computer)

The table below shows selected characteristics of family composition and type of occupation of household guardians. Significance tests were run to check for differences between single and multiple guardian homes when comparing whether the guardian/s have labor or non-labor jobs. Use the information given to answer the questions below.

Household Composition	Guardian Relation	Non-Labor	Labor
Single Guardian Homes	Mother (%)	58.2	36.4
	Father (%)	41.8	63.6
	Total (%)	100.0	100.0
	Number of Families	134.0	129.0
	$z = 3.54$, $p < .001$		
Multiple Guardian Homes	Mother (%)	56.4	38.0
	Father (%)	43.6	62.0
	Total (%)	100.0	100.0
	Number of Families	663.0	403.0
	$z = 5.82$, $p < .001$		

Note: Significance test is difference of proportions, one-tailed.

1. What statistical tests were run?
2. Were one- or two-tailed tests run?
3. What are the sample sizes?

Input the given data into the computer to verify the results shown.

4. Summarize the results of the study making specific references to the number of guardians and labor versus non-labor jobs.

9-7
REDUCING DRUG ABUSE

(Computer)

Assume you read an article that appeared in an education research journal about a study on assertiveness training. The article compared study results looking for the most effective method of reducing drug abuse among youth. Unfortunately, despite huge expenditures of public funds, there is little evidence that any method used has ever reduced consumption of illegal drugs. The results of one study listed below showed significant differences in assertiveness training of the effect of alcohol and marijuana use. Input the following data and analyze the output at the 0.01 level of significance.

Pretest	Posttest
Mean = 2.44	Mean = 3.08
SD = 0.36	SD = 0.46
n = 27	

1. What are the sample sizes?
2. Would you run a right-, left-, or two-tailed test?
3. Are the groups independent or dependent?
4. Does the F test need to be run before the test of dependent means?
5. Can you run a small-sample independent-means test?
6. Compare your results from the independent and dependent means tests.
7. From the results of this hypothesis test, can an inference be made about the effectiveness of assertiveness training?
8. What other factors could have influenced the results of the assertiveness study shown?

10-1
SALT AND BLOOD PRESSURE

Assume you are trying to watch your diet and a friend of yours reminds you that salt is bad for your health. You decide to look up some of the most recent research on salt intake. You find that the majority of recent articles show that salt intake studies compare amount of intake to blood pressure. Also, most of them show reducing salt intake is of little benefit to people with normal blood pressure. Some of the studies showed that people with high blood pressure over the age of 45 could reduce their blood pressure slightly by reducing their salt intake. Answer the following questions about the information given.

---

1. How many variables were studied?
2. What were the variables under study?
3. Are the variables quantitative or qualitative?
4. Are they discrete or continuous?
5. What is the level of measurement for each variable?
6. Would you expect any measurement error?
7. Is it a simple relationship or multiple relationship?
8. Is it a positive relationship or negative relationship?
9. What further information would you like to know about the studies?
10. What are your conclusions about the relationship between salt and high blood pressure?

10-2
STOPPING DISTANCES

(Computer)

In a study on speed control, it was found that the main reasons for regulations were to make traffic flow more efficient and minimize the risk of danger. An area that was focused on in the study was the distance required to completely stop a vehicle at various speeds. Use the following table to answer the questions.

MPH	Braking Distance (Feet)
20	20
30	45
40	81
50	133
60	205
80	411

Assume MPH is going to be used to predict stopping distance.

1. Which of the two variables is the independent variable?

2. Which is the dependent variable?

3. What type of variable is the independent variable?

4. What type of variable is the dependent variable?

5. Construct a scatter-plot diagram of the data.

6. Is there a linear relationship between the two variables?

7. After redrawing the scatter plot, change the distances between the independent variable numbers. Does the relationship look different?

8. Is the relationship positive or negative?

9. Can braking distance be accurately predicted from MPH?

10. List some other variables that effect braking distance.

10-3
MATH CONFIRMS IT

(Computer)

Assume you read an article in a local newspaper about big business wasting money. It states that many businesses waste millions of dollars on office space that goes unused. This loss of money is passed on to the consumer in higher prices for the products or services these companies provide. The Consumer Group Against Corporate Waste (CGACW) collected information on wasted office space and compared that to how overpriced their product or service is. Use the following data to answer the questions below.

Company	Unused Office Space	CGACW Rating
A	32	76
B	125	22
C	73	56
D	243	18
E	45	39
F	68	61
G	112	20
H	38	88
I	99	31
J	45	45
K	101	10
L	70	40
M	281	13
N	133	29
O	50	34
P	69	43
Q	107	37
R	144	21
S	72	80
T	20	90
U	201	30
V	21	55
W	34	56
X	77	30
Y	122	29
Z	55	90

*The CGACW rating is an index on how overpriced a product or service is. The lower the rating, the more the service or product is overpriced.

10-3
MATH CONFIRMS IT
(continued)

In the article, the author states that the more mathematically inclined would understand what the correlation between unused office space and the CGACW rating tells us. He said it is a common-sense conclusion that those companies waste money and pass the debts onto consumers. He also said he was happy to report that math confirms it.

1. Do you agree with these statements?

2. Do you think there is any bias in the CGACW rating?

3. Do you think that having unused office space is an accepted practice since so many companies do it?

4. What does this study confirm (prove)?

5. Are there any other things that may be influencing the correlation coefficient? If yes, list some of them.

6. Compute the linear correlation coefficient.

10-4
STOPPING DISTANCES REVISITED

(Computer)

In a study on speed and braking distance, researchers looked for a method to estimate how fast a person was travelling before an accident by measuring the length of their skid marks. An area that was focused on in the study was the distance required to completely stop a vehicle at various speeds. Use the following table to answer the questions.

MPH	Braking Distance (Feet)
20	20
30	45
40	81
50	133
60	205
80	411

Assume MPH is going to be used to predict stopping distance.

1. Which of the variables is the independent variable?
2. Which is the dependent variable?
3. What type of variable is the independent variable?
4. What type of variable is the dependent variable?
5. Construct a scatter-plot diagram of the data.
6. Is there a strong linear correlation between the two variables?
7. After redrawing the scatter-plot, change the distances between the independent variable numbers. Does the correlation look different?
8. Is the correlation positive or negative?
9. Can braking distance be accurately predicted from MPH?
10. List some other variables that effect braking distance.
11. Compute r.
12. Find the linear regression equation.
13. What does the slope tell you about MPH and Braking Distance? How about the y-intercept?
14. Find Braking Distance when MPH = 45.
15. Find Braking Distance when MPH = 100.
16. Comment on predicting beyond the given data values.

10-5
INTERPRETING SIMPLE LINEAR REGRESSION OUTPUTS

Answer the questions about the following computer-generated information.

Linear correlation coefficient	$r =$	.794556
Coefficient of determination	$=$	.631319
Standard error of estimate	$=$	12.9668
Explained variation	$=$	5182.41
Unexplained variation	$=$	3026.49
Total variation	$=$	8208.90
Equation of regression line	$y' =$	$.725983 X + 16.5523$
Level of significance	$=$	.1
Test statistic	$=$	.794556
Critical value	$=$	.378419

1. Are both variables moving in the same direction?
2. Which number measures the distances from the prediction line to the actual values?
3. Which number is the slope of the regression line?
4. Which number is the y-intercept of the regression line?
5. Which number can be found in a table?
6. Which number is the allowable risk of making a type I error?
7. Which number measures the variation explained by the regression?
8. Which number measures the scatter of points about the regression line?
9. What is the null hypothesis?
10. Which number is compared to the critical value to see if the null hypothesis should be rejected?
11. Should the null hypothesis be rejected?

10-6
MORE MATH MEANS MORE MONEY

In a study to determine a person's yearly income ten years after high school, it was found that the two biggest predictors are number of math courses taken and number of hours worked per week during a person's senior year of high school. The multiple regression equation generated from the study is

$$Y' = 6000 + 4540X_1 + 1290X_2$$

Let X_1 represent the number of mathematics courses taken and X_2 represent hours worked. The correlation between income and mathematics courses is 0.63. The correlation between income and hours worked is 0.84 and the correlation between mathematics courses and hours worked is 0.31. Use this information to answer the following questions.

---

1. What is the dependent variable?

2. What are the independent variables?

3. What are the multiple regression assumptions?

4. Explain what 4540 and 1290 in the equation tells us.

5. What is the predicted income if a person took 8 math classes and worked 20 hours per week during their senior year in high school?

6. What does a multiple correlation coefficient of .77 mean?

7. Compute R^2.

8. Compute the adjusted R^2.

9. Would the equation be considered a good predictor of income?

10. What are your conclusions about the relationship between courses taken, hours worked, and yearly income?

10-7
HORSEPOWER AND PRICE

Assume you are interested in purchasing a new automobile. You first wonder what characteristics about cars are related and second, you wonder which of those characteristics most effects the price of an automobile. Use the computer output for each part to answer the following questions.

Part 1 Questions

1. Which variables show the strongest linear correlation?
2. Which variables show the weakest linear correlation?
3. As price of the cars increases, does miles-per-gallon on the highway increase?
4. Is there a negative linear correlation between turning diameter and horsepower?

Part 2 Questions

5. What is the correlation between price and horsepower?
6. What is the slope of the regression line?
7. What is the y-intercept of the regression line?
8. How much of the variation can be explained by the linear regression?
9. How many cars were in the study?
10. Can the price be accurately predicted over the full range of horsepower?
11. Give three ranges where it is accurate, fairly accurate, and not so accurate.

Part 3 Questions

12. Does the polynomial model fit the points better than the linear model? If yes, how much better? If no, how much worse?
13. What type of polynomial is used for the regression model?
14. What is the polynomial equation?

Summary

15. Summarize the relationship between car prices and horsepower. Make specific references to the outputs. Also, incorporate thoughts about what may influence the relationship.

Part 1 Outputs

Correlations (Pearson)

```
         turndiam   mpg-hwy   horsepow
mpg-hwy   -0.333
horsepow   0.510    -0.618
price      0.367    -0.638     0.723
```

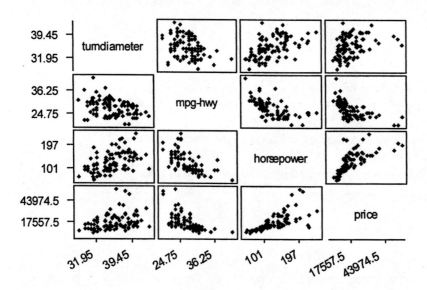

Part 2 Outputs

Regression Analysis

```
The regression equation is
price = -4252 + 168*horsepower

110 cases were used. Seven cases contain missing values.

Predictor         Coef        StDev           T         P
Constant         -4252         2013       -2.11     0.037
horsepow        168.33        15.46       10.89     0.000

S = 6600        R-Sq = 52.3%      R-Sq(adj) = 51.9%
```

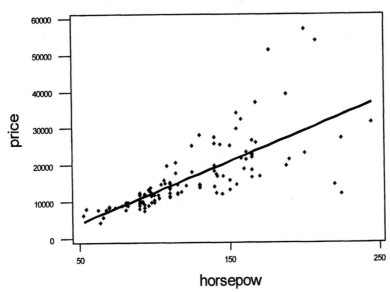

Regression Plot
Y = -4251.92 + 168.332X
R-Sq = 0.523

Part 3 Outputs

Polynomial Regression

```
Y = 21834 - 521X + 557X**2 - .0138X**3
R-Sq = 0.556
```

```
SOURCE       DF    Seq SS         F         P
Linear        1    5.17E+09    118.587         0
Quadratic     1    73201730      1.69139  0.196211
Cubic         1    2.48E+08      5.99577  1.60E-02
```

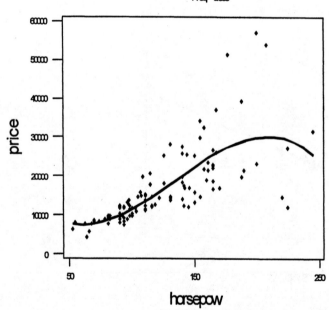

11-1
NEW CAR COLORS

Assume you notice that newer cars on the road seem to be painted colors that were not the same as older models. You decide to collect some data to find out if your suspicions are true. The results of your data collection show that 20% of the new cars are white, 15% are green, 10% are red and 7% are black. The older cars showed 24% are white, 7% are green, 9% are red and 12% are black. It seems that there may be a significant difference between older and newer cars with respect to car color. You decide to run a statistical test. To decide what statistical test is most appropriate, some questions need to be addressed.

---

1. How many variables are under study?

2. What are the variables under study?

3. Are the variables quantitative or qualitative?

4. What is the level of measurement?

5. How many colors are referred to?

6. Create a table organizing the results given above.

7. You read in the newspaper that an automobile paint manufacturer is predicting that purple and gold will be the new "hot" colors. New advances in paint technology will provide more durable finishes with less harm to the environment. Also multi-layer processes will allow the basic colors to be enhanced, such as changing white into pearl. How will this change future studies on car color?

11-2
NEVER THE SAME AMOUNTS

(Computer)

M&M/Mars, the makers of Skittles candies, states that the flavor blend is 20% for each flavor. Skittles is a combination of lemon, lime, orange, strawberry, and grape flavored candies. The following data lists the results of four randomly selected bags of Skittles and their flavor blends. Use the data to answer the questions below.

Bag	Flavor				
	Green	Orange	Red	Purple	Yellow
1	7	20	10	7	14
2	20	5	5	13	17
3	4	16	13	21	4
4	12	9	16	3	17
All	43	50	44	44	52

1. Are the variables quantitative or qualitative?
2. What type of test can be used to compare the observed values to the expected values?
3. Run five separate Chi-square tests, one on each bag and one on the total of all the bags. Compare the results.
4. How does small sample size effect the results of a Chi-Square, Goodnes-of-Fit test?
5. How many skittles were in each of the five tests?
6. What was the degrees of freedom for each test?
7. What would the degrees of freedom be if 1000 Skittles were used overall?

11-3
SATELLITE DISHES IN RESTRICTED AREAS

The Senate is expected to vote on a bill to allow the installation of satellite dishes of any size in deed-restricted areas. The house had passed a similar bill. An opinion poll was taken to see if how a person felt about satellite dish restrictions was related to their age. A chi-square test was run creating the following computer-generated information.

Degrees of freedom ....df = 6
Test statistic ....Chi Square = 61.25
Critical value...............Chi Square = 12.6
P-value.................................... = .06
Significance level ....……….. = .05

	18-29	30-49	50-64	65 & up
For	96 (79.5)	96 (79.5)	90 (79.5)	36 (79.5)
Against	201 (204.75)	189 (204.75)	195 (204.75)	234 (204.75)
Don't Know	3 (15.75)	15 (15.75)	15 (15.75)	30 (15.75)

1. Which number from the output is compared to the significance level to check if the null hypothesis should be rejected?
2. Which number from the output gives the probability of a type-I error that is calculated from your sample data?
3. Was a right-, left-, or two-tailed test run? Why?
4. Can you tell how many rows and columns there were by looking at the degrees of freedom?
5. Does increasing sample size change the degrees of freedom?
6. What are your conclusions? Look at the observed and expected frequencies in the table to draw some of your own specific conclusions about response and age.
7. What would your conclusions be if the level of significance was initially set at 0.10?
8. Does Chi-square tell you which cell's observed and expected frequencies are significantly different?

11-4
CATEGORIZING CONTINUOUS DATA

(Computer)

Use the following data to create two contingency tables.

AGE

Males
16	17	17	19	19	19	18	17	18	17
16	19	19	19	17	16	17	16	19	19
24	31	23	44	21	42	23	43	43	33
30	41	35	40	24	43	22	30	25	32
43	51	55	80	61	58	65	52	67	75
90	63	71	74						

Females
17	16	17	19	19	18	17	19	16	18
19	17	19	17	18	19	19	16	33	23
46	46	23	21	46	47	48	47	48	30
35	24	48	49	47	25	84	54	77	63
51	72	90	57	69	81				

---

1. In the first table, use gender (male and female) as your row variable and age (<20, 20-50, and >50) as your column variable.

2. In the second table, use gender (male and female) as your row variable and age (<18, 18-25, 26-45, and >45) as your column variable. With each table run a Chi-square test of association and print out the results.

3. Compare the results and comment on problems that may occur when categorizing continuous variables.

12-1
COMPUTER-AIDED INSTRUCTION

(Computer)
Assume you have taught statistics to three classes with 12 students in each class. The first group was taught with lectures only. The second group was taught with lectures and group work. The third group was taught with lectures and computer projects. The final grades are listed below.

Group 1	Group 2	Group 3
70	60	75
99	63	80
77	78	45
66	91	79
59	55	84
95	75	50
77	62	76
85	73	35
93	74	91
58	71	35
77	80	82
80	90	15

1. Perform separate t-tests on the three groups (group 1 .vs. group 2, group 1 .vs. group 3, and group 2 .vs. group 3). <u>Note that performing separate t-tests such as this is not an acceptable procedure in this case.</u> Use a 0.10 level of significance. Record your results and draw conclusions about the different teaching methods.

2. Draw conclusions about the validity of the statistical results. Save your results and conclusions to compare to your results from 12-2, One Bad Apple.

12-2
ONE BAD APPLE

(Computer)

Assume you have taught statistics to three classes with 12 students in each class. The first group was taught with lectures only. The second group was taught with lectures and group work. The third group was taught with lectures and computer projects. The final grades are listed below.

Situation 1 Data Sets			Situation 2 Data Sets		
Group 1	Group 2	Group 3	Group 1	Group 2	Group 3
70	60	75	70	60	75
99	63	80	99	63	80
77	78	45	77	78	45
66	91	79	66	91	79
59	55	84	59	55	84
95	75	50	95	75	50
77	62	76	77	62	76
85	73	35	85	73	35
93	74	91	93	74	91
58	71	35	58	71	35
77	80	82	77	80	82
80	90	15	80	90	94

1. Perform a one-way ANOVA on the three groups in situation 1 data sets. Use a 0.10 level of significance. Record your results and draw conclusions about the different teaching methods.

2. Repeat the procedure with situation 2 data sets. Compare the data sets and their respective results. Draw conclusions about the validity of the statistical results.

3. What is the difference between the data sets?

4. Perform separate t-tests on the data sets.

5. Do the ANOVA results match the separate t-test results? Why or why not?

6. Comment on the seemingly conservative results from AVOVA.

12-3
COLORS THAT MAKE YOU SMARTER

(Computer)
The following set of data values were obtained from a study of people's perceptions on how the color of a person's clothing is related to how intelligent the person looks. The subjects rated the person's intelligence on a scale of one to ten. Group 1 subjects were randomly shown people with clothing in shades of blue and gray. Group 2 subjects were randomly shown people with clothing in shades of brown and yellow. Group 3 subjects were randomly shown people with clothing in shades of pink and orange. The results follow.

Group 1	Group 2	Group 3
8	7	4
7	8	9
7	7	6
7	7	7
8	5	9
8	8	8
6	5	5
8	8	8
8	7	7
7	6	5
7	6	4
8	6	5
8	6	4

1. Use ANOVA to test for any significant differences between the means.

2. Do separate t-tests on all possible pairwise comparisons. (Remember that separate t-tests would not be appropriate for this situation)

3. Could the Scheffe' test be used to test all possible pairwise comparisons.

4. Use the Tukey test to test all possible pairwise comparisons.

5. Are there any contradictions in the results?

6. Explain why separate t-tests are not accepted in this situation.

7. When would Tukey's test be preferred over Scheffe's method? Explain.

12-4
TRADITION OR TECHNOLOGY?

(Computer)
A researcher is interested in the influence of test administration modes on student test performance. A random sample of 24 undergraduate statistics students participated in an experimental study designed to test the effects of paper and pencil tests and computer tests. Since computer experience may influence test performance, students were classified on the basis of their previous computer experience. Students were randomly assigned to either a paper-and-pencil test or a computer test. Both tests contained the same test items. The results of the test follow.

Computer Experience	Type of Test	
	Pencil & Paper	**Computer**
Some	45	55
	42	52
	40	58
	37	55
Average	45	53
	42	61
	45	56
	48	58
Above Average	39	72
	34	70
	31	64
	32	66

1. Classify each of the variables.
2. Name an extraneous variable that was controlled in the study.
3. Conduct an ANOVA and construct an ANOVA summary table.
4. State each null hypothesis tested.
5. Indicate the results of each test.
6. Graph the cell means to provide a clear visual display of the results.
7. List the type of interaction if it occurs.
8. What are some of the post-hoc tests that could be used?
9. Write a paragraph on your interpretation of the results.

12-5
AUTOMOBILE SALES TECHNIQUES

The following outputs are from the result of an analysis of how car sales are affected by the experience of the sales person and the type of sales technique used. Experience was broken up into four levels and two different sales techniques were used. Analyze the results and draw conclusions about level of experience with respect to the two different sales techniques and how they affect car sales.

Two-way Analysis of Variance

```
Analysis of Variance for Sales
Source          DF       SS         MS
Experience       3     3414.0     1138.0
Presentation     1        6.0        6.0
Interaction      3      414.0      138.0
Error           16      838.0       52.4
Total           23     4672.0
```

```
                          Individual 95% CI
Experience    Mean    -----+---------+---------+---------+------
1             62.0    (-----*-----)
2             63.0    (-----*-----)
3             78.0                        (-----*-----)
4             91.0                                       (-----*-----)
                      -----+---------+---------+---------+------
                         60.0      70.0      80.0      90.0

                          Individual 95% CI
Presentation  Mean    ------+---------+---------+---------+-----
1             74.0      (------------------*------------------)
2             73.0      (------------------*------------------)
                      ------+---------+---------+---------+-----
```

AUTOMOBILE SALES TECHNIQUES (continued)
12-5

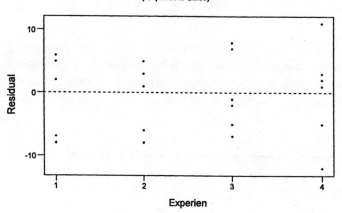

13-1
HIGH SCHOOL CRIMES

(Computer)
Input the district total crimes for each offense listed for the Waltun County Schools in Virginia.

Crimes in Waltun County Schools (1995-1996)

Incident	District Total
Alcohol	26
Arson	6
Battery	68
Breaking and entering	15
Disorderly conduct	122
Drugs	99
Fighting	144
Homicide	1
Kidnapping	1
Theft	51
Sex offenses	5
Sexual battery	4
Sexaul harassment	31
Threats	82
Tobacco	22
Tresspassing	10
Weapon possession	80
Unclassified	21

1. Create a histogram for the district totals. Is it reasonable to assume the data have come from a normally distributed population?

2. Should parametric or non-parametric statistics be used for further analysis?

3. What statistical method could be used to test how "good" the data fit a normal distribution?

4. What are some other non-parametric tests?

5. Are the incident categories clearly mutually exclusive?

13-2
BETTER SAFE THAN SORRY

The medical field continuously conducts research requiring the systematic use of quantitative statistics. Even though the designs used are appropriate and the samples used have been created from random assignment, many studies give conflicting results.

---

1. Is it possible that the statistical techniques are not appropriate or conservative enough for the proposed level of measurement?

2. Could improper use of parametric statistics be a reason for conflicting results in medical studies? Give some examples.

3. What advantages would medical research have by using non-parametric statistics?

4. What disadvantages would medical research have as a result of using non-parametric statistics?

13-3
NO PAIN NO GAIN?

(Computer)
A study was conducted to investigate the effectiveness of a new natural pain medication. First assume the data come from a normally distributed population and run a dependent, 2-sample t-test.

Subject	Pretest	Posttest
1	6.6	6.8
2	6.5	2.4
3	8.9	9.0
4	12.3	7.9
5	11.7	6.3
6	8.4	7.0
7	5.9	2.7
8	12.0	4.1

1. Summarize your results from the first test.

2. Assume the data come from a non-normally distributed population and run a paired sample sign test.

3. Summarize your results from the second test.

4. Compare the results from the two tests. Note there was a large decrease in pain in most subjects.

5. Comment on conditions that may warrant parametric tests even if assumptions are not met.

6. Should both tests be run on any given set of data?

13-4
SIDES EFFECTS FROM MEDICATION

(Computer)
The following table lists the percentage of patients that experienced side effects from a drug used to lower serum blood cholesterol.

Side Effect	Medication	Placebo
Chest pain	4.0	3.1
Rash	4.0	1.1
Nausea	7.0	2.9
Heartburn	5.4	6.9
Fatigue	3.8	2.9
Headache	7.3	3.3
Dizziness	10.0	1.4
Chills	7.0	6.2
Cough	2.6	2.1

1. Input the data and test to see if it may have come from a normally distributed population.

2. What are both sample sizes?

3. Are they large enough to run a Wicoxon Rank Sum test?

4. Are the samples independent or dependent?

5. If the population were normally distributed, which parametric test would you use?

6. Run both the parametric test and its non-parametric counterpart to test for significant differences between the medication and the placebo groups. Compare the results.

7. Briefly summarize each result.

13-5
MEMORIZING AND FAMILIARITY

(Computer)
Assume the follow data were collected from a study conducted on the effect of context on memorizing numbers. It was believed that imbedding numbers in sentences that students were familiar with would help them remember the numbers. The Part 1 column represents the number of numbers memorized without being imbedded in sentences. The Part 2 column represents the number of numbers memorized that were imbedded into familiar sentences.

Subject	Part 1	Part 2
1	2	3
2	6	12
3	5	3
4	5	2
5	0	5
6	5	5
7	5	8
8	2	6
9	3	6
10	2	7
11	4	4
12	3	5
13	7	5
14	3	4
15	9	5
16	3	2
17	6	5
18	3	4
19	9	4
20	1	4

1. What is the purpose of the study?

2. What is the null hypothesis?

Input the data into the computer and answer the following questions.

3. How many degrees of freedom are there?

4. Use a dependent t-test to test for significant differences between the means.

5. Run a Wilcoxon Signed-Rank test.

6. Comment on the differences in the results of the two tests.

13-6
HIGH SCHOOL SUSPENSIONS

(Computer)

The following table lists the reasons and the number of suspensions of boys at a high school in Dunston, CA.

Reason for Suspension (n = 482)

	Nov 1994	Feb 1995	May 1995
Disruption in classroom	14	2	6
Damage of school property	2	2	2
Theft of school property	2	2	0
Assaulting another student	10	10	22
Assaulting school employee	0	0	4
Possession of weapon	0	2	4
Possession of drugs or alcohol	1	4	12
Repeated school violations	55	44	73
Offensive language	28	38	90
Reason unclear	55	13	72

1. Is there a difference in the average number of suspensions at different times of the school year?

2. Run an ANOVA and summarize the results.

3. What assumptions must be met to run an ANOVA?

4. Run a Kruskal-Wallis test.

5. What assumptions must be met for this test?

6. State your conclusions from the Kruskal-Wallis test and compare them to the ANOVA results.

7. What is the difference in what you are testing and how you draw your conclusions for each of the above tests?

13-7
DOES MORE MATH MEAN MORE MONEY?

(Computer)

The following data are from a survey conducted by the American Mathematical Society. The data were analyzed to see if a relationship existed between income and the number of college math credits completed. Input the data into a computer and answer the questions below.

Income $	College Math Credits
12,000	0
17,500	9
22,000	6
45,000	15
31,000	10
15,000	0
20,500	3
65,000	12
94,000	45
18,100	6
75,000	38
64,000	18
27,000	36
87,000	10

1. Compute the Pearson Product Moment Correlation coefficient.

2. Assume a non-parametric test needs to be run. Which one would you run?

3. Run the test and compare your results to the linear correlation coefficient.

4. Will a rank correlation coefficient always be lower than the corresponding Pearson Product Moment Correlation coefficient?

5. Is sample size a concern with correlation?

6. How does it effect rank correlation?

7. What other factors effect correlation?

13-8
PREDICTING RADIOACTIVE WASTE MOVEMENT

Read the following and answer the questions.

Assume you read an article in a local newspaper about the storage of radioactive waste products. In the article it states that errors were made in the prediction of the movement of the waste products after they were stored. It was predicted that it would take 20,000 years for the Plutonium to move one inch. In reality, traces of the stored Plutonium were found one mile away after only five years. Government officials blame faulty equations developed by statisticians. It is argued in the article that the United States needs to use more conservative methods when evaluating radioactive waste product disposal.

1. Do you agree with this statement?
2. Are non-parametric methods more conservative than parametric methods? Explain.
3. Do you agree that more conservative methods are needed in high-risk situations?
4. What are the disadvantages of being too conservative when running statistical tests?
5. Should both parametric and non-parametric methods be run in any analysis? Explain.

14-1
SMOKING BANS AND PROFITS

Assume you are a restaurant owner and are concerned about the recent bans on smoking in public places. Will your business lose money if you do not allow any smoking in your restaurant? You decide to research this question and find two related articles in regional newspapers. The first article states that randomly selected restaurants in Derry, PA, that have completely banned smoking, have lost 25% of their business. In that study, a survey was used and the owners were asked how much business they thought they lost. The survey was conducted by an anonymous group. It was reported in the second article that there had been a modest increase in business among restaurants that banned smoking in that same area. Sales receipts were collected and analyzed against last year's profits. The second survey was conducted by the Restaurants Business Association.

---

1. How has public smoking bans affected restaurant business in Derry, PA?

2. Why do you think the surveys reported conflicting results?

3. Should surveys based on anecdotal responses be allowed to be published?

4. Can the results of a sample be representative of a population and still offer misleading information?

5. How critical is measurement error in survey sampling?

14-2
THE WHITE OR WHEAT BREAD DEBATE

Read the following study and answer the questions.

A baking company selected 36 women, weighing different amounts and randomly assigned them to four different groups. The four groups were white bread only, brown bread only, low-fat white bread only and low-fat brown bread only. Each group could only eat the type of bread assigned to their group. The study lasted for eight weeks. No other changes in any of the women's diets were allowed. A trained evaluator was used to check for any differences in the women's diets. The results showed that there were no differences in weight gain between the groups over the eight-week period.

---

1. Did the researchers use a population or a sample for their study?

2. Based on who conducted this study, would you consider the study to be biased?

3. Which sampling method do you think was used to obtain the original 36 women for the study? (random, systematic, stratified, or clustered)

4. Which sampling method would you use? Why?

5. How would you collect a random sample for this study?

6. Does random assignment help representativeness the same as random selection does? Explain.

14-3
TRADITIONAL FAMILY VALUES

A survey was conducted to find out what people think about traditional family values. The survey was sent through the mail. Read the following survey questions and answer the questions below.

A. You do not believe in abortion do you?
B. Did people need better family values last year?
C. How often do you cheat on your spouse?
D. List three family activities you participate in?
E. Which medium effects family values the most?

_____ TV   _____ newspaper   _____ radio

---

1. Was the survey interview-administered or self-administered?

2. Comment on what is wrong with each question and how each could be improved?

14-4
SIMULATIONS

(Research)

1. Define simulation technique.
2. Have simulation techniques been used for very many years ?
3. Is it cost effective to do simulation testing on some things like airplanes or automobiles?
4. Why might simulation testing be better than real life testing? Give Examples
5. When did physicists develop computer-simulation techniques to study neutrons?
6. When could simulations be misleading or harmful? Give Examples.
7. Could simulations have prevented previous disasters like the Hindenburg or the space-shuttle disaster?
8. What discipline is simulation theory based in?

14-5
MONTE CARLO METHODS

1. Define simulation technique.

2. What are the steps of the Monte Carlo Method?

3. Which steps would be most difficult to complete in realistic situations?

4. Name a real life situation where the steps could be met.

5. Name a real life situation where the steps could not be met. List those steps that could not be met.

6. Use a random number generator to check the following winning numbers from the Florida state lottery, 2, 17, 21, 7, 10. If you would have played the computer generated numbers, would you have won?

7. Is this a Monte Carlo Method? Why or why not?

14-6
SURVEYING AMERICANS

Assume you were reading an article in the March 1996 issue of a popular magazine. It was reported that a poll was conducted by phone to a cross-section of 1045 Americans. It was asked if they thought they were fairly taxed. The results showed that many different groups of Americans responded in very similar proportions.

---

1. Do you think the sample is representative of the American population?

2. What percent of all Americans (roughly 260 million) is the 1045 that were in the sample?

3. What method of sampling would you use to get a cross-section of Americans?

4. List the disadvantages of using systematic samples, cluster samples, stratified samples or a random number generator.

5. How could multistage sampling be done to obtain a representative sample?

6. List the disadvantages of phone surveys.

7. Will your sample still be representative if some of your randomly selected people do not respond to the survey, whatever the reason may be?